PALAEOCLIMATIC RESEARCH AND MODELS

Commission of the European Communities

Palaeoclimatic Research and Models

Report and Proceedings of the Workshop
held in Brussels, December 15-17, 1982

edited by

A. GHAZI

Commission of the European Communities,
Directorate-General Science, Research and Development, Brussels

D. REIDEL PUBLISHING COMPANY

A MEMBER OF THE KLUWER  ACADEMIC PUBLISHERS GROUP

DORDRECHT / BOSTON / LANCASTER

Library of Congress Cataloging in Publication Data
Main entry under title:

Palaeoclimatic research and models.

    At head of title: Commission of the European Communities.
    1. Palaeoclimatology—Congresses.   2. Glaciology—Congresses.
I. Ghazi, A., 1940–   II. Commission of the European
Communities.
QC884.P34   1983   551.6   83–16773
ISBN-13: 978-94-009-7238-4   e-ISBN-13: 978-94-009-7236-0
DOI: 10.1007/ 978-94-009-7236-0

Organization of the Workshop by
Commission of the European Communities,
Directorate-General Science, Research and Development,
Environmental Protection and Climatology Division, Brussels

Publication arrangements by
Commission of the European Communities
Directorate-General Information Market and Innovation, Luxembourg

EUR 8823
© 1983, ECSC, EEC, EAEC, Brussels and Luxembourg
Softcover reprint of the hardcover 1st edition 1983
LEGAL NOTICE
Neither the Commission of the European Communities nor any person acting on behalf of the
Commission is responsible for the use which might be made of the following information.

Published by D. Reidel Publishing Company
P.O. Box 17, 3300 AA Dordrecht, Holland

Sold and distributed in the U.S.A. and Canada
by Kluwer Academic Publishers,
190 Old Derby Street, Hingham, MA 02043, U.S.A.

In all other countries, sold and distributed
by Kluwer Academic Publishers Group,
P.O. Box 322, 3300 AH Dordrecht, Holland

TABLE OF CONTENTS

Session B: Initiation of Glaciation

Session C: Glaciated polar regions and their impact on global climate

FOREWORD

Palaeoclimatology is presently experiencing a period of rapid growth of techniques and concepts. Studies of earth's past climates provide excellent opportunities to examine the interactions between the atmosphere, oceans, cryosphere and the land surfaces. Thus, there is a growing recognition of the need of close collaboration between palaeoclimatologists and the climate modellers.

The workshop "Palaeoclimatic Research and Models (PRaM)" was organized by the Directorate General for Science, Research and Development within the framework of the Climatology Research Programme of the Commission of the European Communities (CEC).

The aim of the workshop was to give to the members of the Contact Group "Climate Models" and "Reconstitution of Past Climates" of the CEC Climatology Research Programme and to some invited scientists the opportunity to discuss problems of mutual interest. About 35 experts from 10 countries took part in the workshop. In general, palaeoclimatologists were asked to identify and discuss the data corresponding to the three topics as defined by the programme committee:

1) Abrupt Climate Changes
2) Initiation of Glaciation
3) Glaciated polar regions and their impact on global climate.

Climate modellers were asked to give their views as to how these specific problems could be modelled, what use could be made of the available palaeoclimatic data and which complementary data are needed for modelling.

This volume contains the outcome of the workshop i.e. a report and the summaries of the presentations in the form of proceedings. The workshop report was prepared mainly by rapporteurs in collaboration with the Programme Committee (see appendix) and was later edited to make it more cogent and lucid without changing its scientific content.

This volume contains a fairly complete account, particularly from the European viewpoint, of the status of research and research needs in the areas, which were discussed.

Thanks are due to Dr. J.C. Duplessy for the suggestion of holding this workshop and Dr. Ph. Bourdeau, Dr. H. Ott and Dr. R. Fantechi (DG XII) for their support in organizing the meeting.

Brussels, July 1983

A. GHAZI

WORKSHOP REPORT

Session A : ABRUPT CLIMATE CHANGES

A.1. INTRODUCTION

Abrupt climatic changes have been revealed by many continental and
marine palaeoclimatic records. However, coring disturbances, dating
uncertainties, sedimentary hiatuses and particular problems related
to the indicators of past climates impede estimation of the time span
of such abrupt climatic transitions.

From the existing literature no clear definition of "abrupt changes"
could be found. As a working definition the following can be proposed : an abrupt climatic change has a time-scale of the order
between 50 and about 200 years, while the temperature difference is
of the order of about half the difference between glacial and interglacial, i.e. 2-3 K.

Climatologists expect to learn from past climate studies whether
climatic conditions could change drastically within times of the
order of a century or less. Therefore, the available proxy-indicators
of past climates are reviewed, examples of abrupt changes are given
and the related problems are discussed here. Finally, some recom-
mendations to estimate more accurately the duration of climatic transitions are formulated.

A.2. INDICATORS OF PAST CLIMATES

Ice cores from the polar regions contain a wealth of information on
past environmental conditions, including climate. Along with the
snow, all kinds of fall-out from the atmosphere are deposited on the
ice sheet surface. In the coldest areas, no melting occurs and all
of the impurities, including samples of the atmospheric air, remain
in the ice in unchanged concentrations and in an undisturbed sequence,
as the annual layers sink into the ice sheet under continuous pressure.
Close to the bedrock, the annual layer thickness approaches zero.
An ice core drilled to bedrock in a favourable location contains a continuous sequence of hundreds of thousand annual
layers.

This makes ice cores a rich source of information on the environmental system and its mechanism. Long time series of geophysical
and geochemical parameters are being measured, such as :

- oxygen isotope ratios, that are indicative of surface temperature
  at the time of deposition;

- CO_2 concentration in the atmosphre;

- concentration of polar continental dust that is indicative of the
  turbidity of the polar atmosphere;

- concentration of strong acids and sulfates that are indicative of
  volcanic activity;

- NaCl that originates from sea spray and that may be indicative of
  storminess and/or ice-cover.

On the continent, dust transported in the air and weathering products accumulate locally, for instance in lakes and river. Most sediment results from the organic productivity of lakes and erosion of organic material from adjoining upland. The geochemical and sedimentological study of these deposits provides information on the climatic conditions which prevailed at the time of their deposition. However, the most promising way to obtain quantitative informations on the climatic conditions over the continents at our latitudes during the last 150,000 years is the analysis of pollen deposited in lake sediment and peat bogs. The determination of this pollen is used to reconstruct the past vegetation and to infer the climatic conditions compatible with the reconstructed vegetation cover.

Cores from the ocean floors contain material deposited at a fairly constant rate, something of the order of a few cm per thousand years. This material is constituted essentially from weathering products originating from the continent (clay) and from carbonate or silica shells of organisms (foraminefera, coccoliths, diatom, ...) that have lived in the overlying sea water. The variations of the oxygen isotopic composition of these shells reflect those of the volume of ice stored over the continent and have provided a worldwide stratigraphic framework, which can be used in all the oceans. Moreover, the composition of the fauna in ocean surface water depends mainly on the temperature and, hence, if the composition of the fossil fauna is determined by counting, it may be transformed into an estimate of the surface palaeotemperature by statistical analyses using empirical transfer functions. The standard error of such estimates is about 1.5° C.

As a consequence, deep sea records have proved capable of providing a stable chronological framework for the evolution of past climates during the last million years. More detailed climatic records can be obtained in places where the accumulation of sediment is very high (up to 50 cm per thousand years), but the length of obtainable deep sea core is technically limited to about 20 meters and the time interval covered by such cores is reduced to a few tens of thousand years. Deep sea records help to improce the reconstruction of continental climatic fluctuations, especially with regard to coastal areas.

A.3.   EXAMPLES OF ABRUPT CLIMATIC CHANGES

Climatic crises of brief, but undetermined duration have been described in the Sahara, where sedimentary sequences were interrupted by periods of drying and erosion. Severe droughts, interrupting lake sedimentation, appeared to have occurred between 8,000 and 7,000 years B.P., i.e. before a general drying of the Sahara.

18,000 years ago, the high latitudes of the northern hemisphere were covered with large ice sheets, which developed over northern America, Scandinavia and northern Europe. The deglaciation, which started about 15,000 years ago, was not a slow process and may have been a succession of sudden events. At the same time, strong temperature changes were observed in the ocean. Within about 2000 years or less, the sea surface temperature along the Atlantic coast of Europe increased by some 14° C in the Bay of Biscay and temperatures were at least as warm as those of today. This complex warm phase, which includes the continental Bölling and Alleröd interstadials, was interrupted by a major spread of polar waters and in less than 1000 years (perhaps less than one century), marine temperatures were almost as low as those of the last glacial maximum. During this event, called Younger Dryas, glaciers readvanced in northern Europe. Vegetational changes in southern France and in Florida, probably in response to climate as early as 15,000 B.P. are correlated with the complex deglaciation process.

The Last Interglacial, known as Eemian, lasted only some ten thousand years, from 125,000 to 115,000 years B.P. The Eemian, which has been studied using ice cores, deep sea cores and continental deposits, is so similar to the present interglacial that a detailed study of its late phase and its termination is of great interet, the more so as the ice and deep sea cores contain evidences of abrupt climatic changes not only at the end of glacial conditions but most probably also at the end of the interglacial. However, the dating has errors of as much as a few thousand years for sediments of this age and it is not possible to estimate the duration of these transitions accurately.

A.4. THE PROBLEMS

It was observed that the standard forms of presentation of the pollen diagrams may give a false impression of sudden climatic change because of the apparently abrupt appearance of new taxa. These abrupts changes may be real, but apparently abrupt appearance of new taxa may arise from a hiatus in the sedimentation or from the dynamics of the vegetation itself. Calculations of the time required for such changes, taking into account variable sedimentation rates and variable compression indicate that in some cases 500 to 1000 years are required for a major change to take place. This is explained by the need of a migrating species to travel from one geographic region to another and to establish its population in a new region. The phenomenon of migration is a constant climate may give a spurious impression of climate change at one site. The need to distinguish between migration and vegetation response on one hand and to select particularly sensitive sites with a good sedimentary record on the other appear decisive to interprete the observed abrupt changes in the pollen content of continental deposits.

In the ocean, all kinds of sedimentary records may be disturbed by mechanical mixing of the upper few centimeters of the sediment due to the activity of benthic organismes that live in the oxygenrich bottom water. The consequent temporal smoothing of the climatic records is of course most serious in areas of low sedimentation rate, and complications can occur if a noticeable productivity change is correlated with the climate change.

The acquisition of high quality ice cores is difficult. The Greenland site of Camp Century was a poor drilling site, because this area may very well have been exposed to higher altitude changes more than most other areas in Greenland. This is one of the reasons why the Camp Century isotopic record has never been provided with a temperature scale. The new deep ice core from South Greenland (Dye 3) was drilled in an area with some summer melting, making its dating difficult even for the recent past. Furthermore, it reaches only 90,000 years back in time in a continuous sequence, which means that the last interglacial is not adequately represented in this core. Moreover, an experimental dating of the ice is at present only possible by counting annual layers. Even if there is no summer melting, these layers disappear by diffusion in the ice older than 10,000 years and the dating of ice cores is possible only by models.

A.5. RECOMMENDATIONS

During the U.S.-Danish-Swiss joint effort, Greenland Ice Sheet Program, it became obvious that Central Greenland is, scientifically, the most favourable location for deep ice core drilling because :

- the bedrock topography is smooth, which ensures simple ice flow modelling;

- the ice is frozen to the bedrock, which ensures a very long time range of a deep ice core (possibly more than a million years).

- The accumulation rate is high enough to allow absolute dating at least back to the termination of the last glaciation, to ensure a continuous layer sequence throughout the last glacition and, most likely, through several preceeding ones.

- No melting occurs at the surface, which means that the air trapped in the bubbles of the ice has the same composition as the atmosphere at the time of deposition.

Hence, a new deep ice core should be drilled in Central Greenland.

At lower latitudes, priority should be given to the study of continuous sequences in which a high resolution climatic record can be obtained. Attention was drawn to the advantages of studies of lake sediments with annual laminae when exact chronological control was required and to the possibility of obtaining high resolution marine records by coring in the continental margin, where detrital input is high.

In the effort to choose periods which were critical to illustrate sudden climatic changes, it was agreed that the Younger Dryas period was the best example of such a change. The new development of the carbon-14 dating method linked to the use of accelerators should be applied to determine the precise chronology of this event, both in ice core and in continental or marine sediments. A collaboration between geologists, physicists and biologists seems most appropriate to simulate an interdisciplinary study of these samples and to avoid the pitfalls described above.

Opportunities to link studies in palynology with other palaeoclimatic studies are specially strong in relation to marine core and ice core studies, as well as various physical measurements which provide evidence of past climatic environments. This suggests that emphasis might be in future on :

- Long continuous records, especially where good dating possibilities exist. These might be on the Atlantic coast because west european climate is largely determined by the ocean and this is the best region to link ocean and land records. Another possibility could be the Mediterranean, because it is still much less known than northern Europe and offers good chances to obtain long records.

- Technical improvements, especially in understanding the significance of the data obtained from fresh-water sediments. Studies of
pollen influx after the best possibilities at presend and the extensive use of surface samples to establish transfer functions
providing estimates of past climates could allow for significan
advances this field.

Since the two warming events (mentioned under A.3.) are correlated
with an abrupt increase of atmospheric CO_2, the hypothesis of a
feedback effect : warming → weaker Hadley circulation → prevailing
downwelling and rapid increase of CO_2 and H_2O from equatorial oceans
→ further warming (see FLOHN, this volume) should be investigated.
A reversed feedback mechanism could occur after a glacial iceberg-
meltwater surge causing colling and stronger circulation → prevailing
upwelling   decreased atmospheric CO_2 and H_2O.

There is a great need to apply methodologies such as transfer functions
and other mathematical quantitative models used in ice- and deep-sea
core and land record studies. Utilization of the geological date e.g.
evidence of climatic change from palaeosoils, sedimentation pattern
and landscape surfaces could be useful to complement information,
yielded from other studies about past atmospheric conditions acting
on the lithosphere and the biosphere.

Session B : INITIATION OF GLACIATION

B.1. BACKGROUND

Glacial-interglacial alternations have been characteristic of the past
2.5 million years of earth's history. By obtaining continuous records
of past temperatures and ice volumes, primarily from deep sea sediments, scientists have established that probably the primary cause of these
quasi-cyclic alternations is changes in the earth's orbital geometry.
This mechanism was first formulated in a clear way by M. Milankovich.
Thus one aspect of any study of the initiation of glaciation must be
the investigation of this mechanism in more detail. However, the
particular situation in which the earth had no ice on North America
or Fennoscandia poses additional problems, especially in view of the
evidence that ice accumulation was very rapid. Thus we hope that a
considerable amount may be learned by the careful study of the evolution of climate about 120,000 years ago; this is the last time that
the earth went through this crucial phase. It seems likely that a
focus on this particular episode will be the most fruitful research
area for furthering our understanding of the initiation of glaciation.
However, we also have to consider whether it is valid to assume that
the manner in which climate responds to orbital forcing has been sufficiently consistent that we can afford to take only one glacial
inception as a model, or whether other predictable or unpredictable
factors intervene. This requires that we continue attempts to model
all aspects of the glacial-interglacial record, and to evaluate model
output by careful reference to the geological record.

B.1.a. The orbital forcing

Until quite recently only quite naive interpretations were made of
the insolation variations deriving from orbital variations; meteorolo-
gists tended to assume that their effects would be trivially small.
However, if one looks at particular months and particular latitudes
one can find insolation variations up to 30 % ! There is still scope
for more formulations of the orbital effects : in some contexts insolation gradients are important. For some systems (plants) insolation
hours may be important, and so on.

B.1.b. The climate response

Very exciting work is in progress modelling climate parameters as a
function of orbital forcing; atmosphéric circulation, ice-sheet growth,
sea ice cover, snow cover. These studies must continue; interaction
between the groups should be stimulated. Communication with geologists
is very important, so that modelers take account of new information on
the boundary conditions and their changes; for example, the recent
information that during the last glacial maximum the atmospheric concentration of CO_2 was about half of its present value.

B.2. THE LAST INTERGLACIAL TO GLACIAL TRANSITION

The Last Interglacial (referred to as isotopic substage 5e in the oceanic record and as the Eemian over the european continent) lasted about 10,000 years, starting about 125,000 years ago and ending about 115,000 from now. Its end was marked by a new phase of ice-accumulation over the continents, which culminated about 110,000 years ago. At that time, the sea level had dropped by about 60 meters and large ice caps covered Northern America and Northern Europe. The occurrence of these large volume of ice, strongly depleted in oxygen-18, resulted in a worldwide increase of the ocean water $^{18}O/^{16}O$, which was well reflected in the oxygen isotopic composition of planktonic and benthic foraminifera. Consequently, this phase is referred in the oceanic record as isotopic substage 5d.

The transition between isotopic substages 5e and 5d should be studied both by modelling and by a very precise search of the climatic para-meters which may have changed first during this climatic transition.

Modelling of this transition should use both orbital data and boun-dary conditions of this epoch. However some of them are still poorly known, such as the continental albedo and the sea-ice extent. Special attention should be paid to the following parameters :

a) Sea-ice changes during the 5e/5d transition may be important in Labrador Sea area and in the Norwegian Sea.

b) The climatic evolution of Antarctica and of the Southern Ocean arond Antarctica is still poorly understood and should be studied in great detail. For example, some faunal and isotopic data sug-gest that this area was already much colder than now at the end of isotopic substage 5e, before the initiation of the ice-growth phase over the northern hemisphere continents. Such a behaviour of the southern hemisphere climate should be checked carefully.

c) The ice-accumulation rates are still unexplained. Isotopic and continental field evidence indicate that the ice accumulation during the intiation of the glaciation was a rapid process. Less than ten thousand years have been necessary for the 60 meters sea level drop and the advance of continental glaciers. Such a heavy accumulation of snow and ice over the high latitude continents requires snow-falls at least four times heavier than those known today and a modelling effort should be made to find a realistic mechanism able to produce them.

d) We need to have a much better understanding of the vegetation and of the continental climate as ice began to accumulate. Was it already changing ? How fast ? Although free migration rates confuse the interpretation of the vegetational succession linked to a glacial to interglacial transition and to the following inter-glacial to glacial transition, it should be possible to develop transfer functions to describe the last part of an interglacial. These transfer functions should be applyed to the main pollen se-quences such as Grande Pile, les Echets or Padul and perhaps to annually banded diatomites.

e) The atmospheric concentration of CO_2 is an important aspect of the atmospheric heat budget, because CO_2 is transparent to visible radiation, but absorbs in the infra-red region of the radiation emitted by the earth's surface. It therefore contributes appreciably to heat the atmosphere (greenhouse effect). Ice core studies have shown that the atmospheric concentration of CO_2 was about half of its present value during the last glacial maximum. However we do not know when and why the atmospheric CO_2 concentration began to change. If it would appear that the drop in CO_2-concentration leads the climatic change, then this variation could be a cause of the glaciation by reducing the greenhouse effect. If, more probably, the CO_2 concentration lags behind the climatic change, then this CO_2-variation could be a response of the global carbon dioxide cycle to the climate. Consequently, the resulting decrease in greenhouse effect would amplify the glaciation (see also A.5.).

B.3.   LONGTERM RESEARCH NEEDS

We need ice sheet models able to explain the evolution of ice sheets under various accumulation rate conditions. Since oxygen isotopes provide a detailed time-record of the volume of ice stored over the continents during the last million years, the comparison of the model output with observational records is most valuable. This modelling effort must focus on the relationship with real record, rather than on the development of esoteric models that produce output that is visually similar to the geological record, but do not show appropriate coherency with the orbital forcing. Then, we need to know whether a long (several hundred-thousand years) geological record can be simulated by a model with specified parameters, or whether one has to consider a certain amount of parameter-changing, which would reduce the predictive value of such models.

Session C : Glaciated Polar Regions and their Impact on Global Climate

C.1.   THE PROBLEM

Evidence for very cold climatic conditions have occured several times during the Phanerozoic. Because of specific plate-tectonic settings which resulted in polar positions of either small ocean basins or continents, the Paleozoic spells of cold climates led to unipolar glaciations. Only during the Cenozoic a plate-tectonic situation evolved which led to ice shields on Antarctica and around a possibly ice covered Arctic Ocean.

Flohn reported on the important problem of the climatic asymmetry between the southern and the northern hemisphere which seemed to characterise the early history of Cenozoic glaciations. The volume of the Antarctic ice-shield seems to have fluctuated considerably through time, but it is generally believed that it was a persistent feature since Miocene times at least. The earliest indications for ice in Antarctica can be traced back to the Oligocene, to a time when plate-tectonic movements led to the separation of Antarctica and Australia.

The history of glaciation on the northern hemisphere is considerably less well known, but it is believed that the ice shileds which developed on the continents around the Arctic Ocean and of the Arctic sea ice cover have been intermittent features only. Our lack of ability to document the history of northern hemisphere glaciation with data from the Arctic Ocean (because of the virtual unavailability of samples of pre-Pliocene age), from the Norwegian-Greenland Sea and from the Bering Sea as well as from the adjacent land areas in the real obstacle for understanding this development.

Thiede explained that late Cenozoic sediments from the bottom of the presently ice-covered Arctic Ocean have up to now eluded an easy explanation of their depositional environments. They contain few or only sporadic fossils and they seem to accumulate at very solw rates $(0.5 - 2 \ mm:10^{-3}y)$. In general, they consist of fine-grained grayish terrigenous muds which contain varying proportions of coarse-grained ice-rafted detrital material. Hypotheses about the history of the Late Cenozoic Arctic Ocean palaenoenvironements therefore range from a permanent ice cover, to intermittent and/or loose sea-ice cover, to long ice-free stages, despite   evidence of long spells of cold glacial climates from the surrounding continents.

Two aspects of Late Cenozoic Arctic Ocean depositional environments are of particular interest :

1. Fossiliferous intervals in the sediment cores are marked by a dramatic increase of planktonic as well as benthic fossils (both
   calcareous and non-calcareous ones). Especially calcareous fossils
   (planktonic and benthic foraminifers, ostracods, molluscs, echinoderms, etc.) are well preserved, even in samples from abyssal water
   depths. These intervals can be correlated over wide distances and
   seem to indicate times of high plankton productivity in near-
   surface waters, and the respons of higher standing stocks of
   benthic organisms. The fossiliferous zones coincide or overlap
   sometimes, but not always with times of intensive ice-rafting.

2. Ice-rafted material has been found in Upper Miocene to Recent
   sediments from the Arctic Ocean. It occurs as pebbles in a clayey
   matrix throughout the sediment column, but it is enriched in a
   number of correlatable horizons (6 or 7) which seem to point or
   phases of intensified ice-rafting. The ice-rafted components
   consist of broken to subrounded fragments of metamorphic,
   crystalline and sedimentary rocks with diameters of 0.5-80 mm.
   Because of the poor stratigraphic resolution of the available
   cores it is presently not clear which type of palaeoclimatic setting led to the important, but intermittent influx of ice-rafted
   material approximately at the same time cover the entire Arctic
   Ocean.

From Flohn's and Thiede's presentation it was clear that many open
problems remained : How and when are polar regions glaciated ?
What is the impact of plate-tectonics and of resulting changes of
ocean circulation ?

C.2.   MODEL STUDIES

A summary was given of the numerous atmospheric general circulation
model studies which have included a change in sea ice distribution.
In three experiments, completely removing the Arctic sea ice produced a reduction in polar surface pressure, and weakened the nothern
hemisphere mid latitude westerlies. Other studies have been made
in which the sea ice boundary has been altered in one or both
hemispheres. In three out of four integrations in which the sea ice
boundary was moved north (or south), the neighbouring depression
tracks also moved north (or south). This was generally the case in
two other integrations in which factors other than sea ice were also
changed (sea surface temperatures, land ice distribution) to make
them consistent with the glacial maximum 18000 years ago. The
results of these integrations need to be interpreted cautiously in
view of known deficiencies in model simulations in high latitudes,
and, in the case of the integrations in which larger areas of sea
ice were removed, the unrealistic nature of the change in the model.

C.3.   COMMENTS AND RECOMMENDATIONS

Geologic-palaeoclimatic-palaeoceanographic studies have to be carried
out to help describe mainly the northern hemisphere glaciation.

1) New samples will have to be collected in the Arctic Ocean in
   attempts to document the early history of the Cenozoic climatic
   deterioration. Very serious shortcomings exist in our dating
   techniques and in the low sedimentation rates observed in Arctic
   Ocean sediment cores; it is presently impossible to time the
   various changes of the depositional environment which were recorded
   in the sediment cores. Special emphasis should be given to Eocene
   and Miocene sediments.

2) It is presently quite unresolved how the Arctic Ocean responded
   to the late Tertiary and Quaternary northern hemisphere climatic
   changes. What is the significance of the horizons enriched in
   the ice-rafted material and in biogenic components ? Was the
   Arctic Ocean permanently or only intermittently covered by pack-
   ice, or even by ice-shelves ?

Atmospheric general circulation models may be used in at least two ways to help increase our understanding of palaeoclimates.

(1) Sensitivity Studies. One model boundary condition (e.g. sea ice extent, sea surface temperature) is changed between a control and anomaly integration to determine the sensitivity of the model circulation to that change. The change need not be realistic, but should be large enough to produce changes easily detectable above the model's natural variability. Removing the Arctic sea ice (Hills, this volume) is one example of a sensitivity experiment; changing solar insolation is another (Royer, this volume).

(2) Coupled to general circulation models of the oceans and sea ice. The usefulness of general circulation models of the atmosphere is limited by the need to prescribe sea ice extents and sea surface temperatures. The processes which should be included in a numerical model of sea ice were reviewed. Some of these processes have been included in general circulation models in which the atmosphere, ocean and sea ice are coupled interactively. However, such coupled general circulation models are expensive to run, and are at an early stage of development. There are only two coupled model integrations made with realistic geography and a full seasonal cycle for present day conditions which have been published in the open literature. Hence it is not surprising that they have not yet been used for studies of palaeoclimates. Unfortunately, simulations made with couple models are poor in high latitudes (e.g. the Southern ocean is warmer and the Antarctic sea ice less extensive than observed). Palaeoclimatologists have expressed interest in the advance and retreat of the North Atlantic oceanic polar front over the last the thousand years or so. It has been speculated that the advances are triggered by the release of large numbers of icebergs from neighbouring ice sheets. It is unlikely that this phenomenon can be investigated using numerical models in the near future. Palaeoclimatologists would have to provide modellers with the appropriate boundary conditions, and in particular, estimates of the rate of the release of ice.

Appendix to Session Reports

Relationships between Modelling and data interpretation

Climate Models have been calibrated and tentatively validated using historical and geo-ecological data namely :

- to understand the mechanisms which are responsible for climatic variations and variability;

- to study the sensitivity of the climate system to different internal para-meters, parameterization of the physical processes, feedback mechanisms and to the external forcing;

- for the interpretation of reconstructed past climates and their palaeo-geographical implications;

- for the simulation of some particular past climates, climatic variations and climate evolution at different time scales : palaeogeography and Tertiary versus Quaternary climates, astronomically induced glacial-interglacial cycles and higher frequencies of the climate spectrum related to internal mechanisms, e.g. cloudiness and oceans, or to man's impacts (such as CO_2, trace gases, deforestation).

- in order to study the transient response of the climate system to a parti-cular forcing.

There is a large variety of atmospheric models which are used for both pre-diction on time scales from one month to a season and for climate sensitivity studies for time periods from a season to decades, thousands and hundreds of thousands of years. They can be distinguished among themselves by the way of treatment of the major physical processes responsible for the evolution of the weather and the climate. Although substantial progress has been achieved in the development of the comprehensive atmospheric models, the simplified climate models have been used so far as an important tool for the study of various properties of the global climate system. In particular, they are considered as an important addition to GCMs for studying certain feedback mechanisms of the global climate system (e.g. ocean-climate), evaluation of climate sensitivity to individual external forcings of anthropogenic (e.g. CO_2 deforestation) and natural (e.g. astronomical forcing) origin as well as the interpretation of the causes of the changes of the global climate simulated by GCMs on decadal and longer time scales. They are also important and unique for the simulation of the dynamic behaviour of the climate system, namely during the abrupt climate changes, and the transient response of the climate system to forcings such as CO_2-change. One of the best ways of identifying regional (not local) climatic changes with potentially good signal-to-noise ratios would thus be to search for these in the results of detailed time-dependant simulations performed with a coupled atmosphere/ocean model using realistic geography and plausible CO_2 increase scenarios. The problem of estimating the transient climate response also has a bearing on the selection of precursor indicators of a CO_2 warming.

There are mainly two ways in which General Circulation Models can be applied to palaeoclimatic research :

- global reconstructions of the atmospheric circulation for specific times of the past, when the corresponding surface boundary conditions (SST, albedo, ice extent, etc.) have been reconstructed globally from various palaeoclimatic data (deep-sea and ice cores, paleus, geological data, etc.);

- sensitivity studies of the impact on the model simulations of the prescribed variations of particular external factors, in order to analyse the mechanisms of the interactions of these factors with climate variations. Removing the Arctic sea ice and changing solar insolation are two examples of such studies.

The possibilities of a valuable cooperation between palaeoclimatologists and modelers has already been demonstrated in the CLIMAP Project, by palaeoclimatic reconstructions of the summer and winter surface conditions for the last glacial maximum 18,000 years B.P. and numerical simulations of the corresponding atmospheric circulation.

Other potentially interesting periods, where variations of the external insolation forcing play, probably a more significant role, are the optimum of the last (Eemian) interglacial and the subsequent rapid transition to more glacial conditions around 115,000 B.P. The sea surface temperature anomalies for both periods have been recently reconstructed and could be used as impact for GCM simulations.

For a complete reconstruction, further data on sea ice extent and land albedo will be necessary.

The interpretation of the model results needs caution, because random fluctuations are not completely averaged out with a simulation sample necessarily very limited in time, and because the models themselves can have some unknown bias resulting from various approximations in the physical parameterizations.

Therefore a comparison of the sensitivity of different CGMs for the same experiment could be very valuable to assess the reliability of their results in palaeoclimatic simulations. There are at least 4 or 5 GCMs in European countries that could possibly contribute in a comparative simulation. Such an experiment should be defined in close association with palaeoclimatologists in order to bring together as much data as possible for model initialization, validation or interpretation of the results.

WORKSHOP PROCEEDINGS

REVIEWS

Actual Palaeoclimatic Problems from a Climatologist's Viewpoint

H. Flohn (University of Bonn)

Introduction

One of the great challenges of our time is the possibility of a major
climatic change as initiated by man's activity. Changing of atmospheric
composition (CO_2, N_2O, CH_4, Chlorofluoromethanes) and of its particle content,
conversion of the soil/air interface (land-use, deforestation, irrigation)
and other stresses increasingly modify the atmosphere, its boundary conditions and the internal feedbacks of the climatic system. Modelling of this
climatic system has lead to remarkable achievements, but has also left
important gaps to be filled. Up to now no model has simulated in sufficient
detail the present surface climate : distribution ans seasonal cycle of precipitation, of snow-cover and polar drift-ice, extension and impact of
oceanic upwelling and other essential features are only incompletely understood. A systematic review of climatic models has been given by Schlesinger
(1982).

At present we are unable to outline sufficiently realistic perspectives of
future climates, the more so since we do not understand the causes of past
climatic changes. New techniques and new facts are substantially en-
larging our views; such results force us continuously to reconsider our
hypotheses and ideas. The purpose of this personal review is to bring some
of these new problems to the attention of model designers; nature itself
can find - on-line and within her new time-scale - quite different solutions
for the complete set of unabbreviated and unparameterized equations under
different boundary conditions.

Little Ice Age (1570-1860)

During this period (Lamb 1977, 1982, Pfister 1983) several cold episodes
lasting not more than a few decades lead to a hemispheric advance of mountain
glaciers to a level identical with the highest level ever reached during the
last 9000 years, i.e. during the Holocene (Gamper et al. (1982)). During its
coldest episodes (e.g. 1683-1700, 1812-1850) individual seasons, years or
clusters of years occurred which would now provoke journalists to announce
the immediate onset of an ice-age. However, other seasons and years were
warmer than the hottest years of the last decades - all kinds of weather
extremes were more intense and more frequet than now. Longer warm periodes
(e.g. 1902-30) interrupted this cold period, which also had its forerunners
- notably between 1310 and 1340, as well as 1428-65. The extension of Arctic
drift-ice probably has varied about 2×10^6 km^2 between a minimum during the
Viking colonisation of Greenland, when the East Greenland Current was ice-
free (10th-13th century), and its maximum around 1695.

Lamb has published several weather anomaly maps for characteristic seasons;
18th century wind statistics from London, Amsterdam, Copenhagen and Berlin
are waiting to be analised in detail. The weather map series 1781-1786
(Kington (1975)) demonstrate a unusually high frequency of meridional circulation patterns with blocking anticyclones - even higher than our experiences during the last 20 years. The most striking event in the 1690's
was the advance of polar water masses with seasonal ice blocking Iceland
and reaching the Faeroes and perhaps even western Norway, with sea surface
temperatures (SST) 3-5° C lower than now at the entrance of the North Sea

(Lamb 1979). This caused prolonged winters and a strong cooling of spring and early summer in northern and central Europe (Fig. 1). Simultaneously the rainy season in the Mediterranean, northern Africa and the Near East became longer and more intense; the probability of droughts in the Sahel belt (Nicholson 1981) and India (Mooley-Pant 1981) was apparently higher.

One of the characteristics of this period is the clustering of years with similar anomaly patterns - as in our present period. The cold winters 1939-42, the series of mild European winters in the 1970's, the three intense iceberg seasons near Newfounland 1971-73, the sequence of very harsh winters 1976-79 in eastern and central USA are good examples. One of the most unusual cases of the past are the cold European summers between 1813 and 1817, including the famous 1816 ("year without summer") event. During the coldest decades the annual temperatures were about 1.5° C lower than now, and the vegetation period 4-6 weeks shorter (Fig. 1). Disastrous impacts on food and economy accompanied these sequences of anomalous years.

Without discussing the possible causes of the "Little Ice Age" the following facts are essential :

1) Variable frequency of zonal ("high index") and meridional ("low index") circulation patterns, including blocking anticyclones.

2) Clustering of years with "anomalous", mostly meridional circulation patterns.

3) Rearrangement of North Atlantic wind-driven ocean currents (including the Gulf Stream system) and water masses, with marked changes of SST and sea-ice.

Since the available evidence originates mainly from Europe, eastern North America and northern Africa, the Atlantic section seems to be stronger affected than the Pacitif section including Eastern Asia. But this conjecture is by no means verified and awaits further investigation with the admirable historical sources of China and Japan.

Late Glacial and Early Holocene (14 ka - 6 ka BP)[1]

Palaeoclimatic evidence from the period of the decay of the great ice-sheet of the last glaciation is abundant, at least in Europe, Africa and North America. Nevertheless, unsolved controversies still exist, partly due to the variable relation between the chronologcal age (e.g. from the tree rings) and the C_{14} dated "radiocarbon-age" with its "wiggles" (De Jong 1979). Due to the quite different volume of ice-sheets and their decay-time, the warmest period ("optimum") of the Holocene is diachronous. While the seasonal Subantarctic drift-ice can respond quite fast (optimum at or before 10 ka), the Scandinavian ice-sheet did not disappear before about 8.5 ka, the Laurentide ice-sheet split (about 7.8 ka ago) by an incursion of the sea into Hudson Bay, while its residues in Labrador/Keewatin remained until 6 or 5 ka. Thus in the Subarctic the optimum may have happened as late as 4.5 ka ago, while in Europe the warmest period occurred around 6 ka. Some remnants (Baffin Island, Ellesmere Island) remained until now.

1 - 1 ka (kiloanno) = 10^3 years, 1 Ma (Megaanno) = 10^6 years, BP = before present. The use of such terms (as in Physics) should be recommended; others are often misleading or linguistically incorrect. Some remnants

During this period of ice decay one of the most important orbital parameters changed significantly : it is the time of the perihelion (shortest distance sun-earth). Now it occurs in early January, thus giving higher solar radiation to the southern hemisphere. 12-8 ka ago the perihelion occurred during northern summer and favoured the northern hemisphere; the extraterrestrial summer radiation was then about 7 percent higher (cfr Berger, Kutzbach). This change was mainly effective in temperate and subtropical continental areas in sufficient distance from the ice-sheet.

The alpine glaciers had retreated substantially as early as 14 ka; between about 12.8 and 10.8 ka a complex warm period (Bölling-Alleröd, with only weak interruption) occurred, which was abruptly terminated by a marked cold-dry phase (Younger Dryas) with its peak around 10.5 ka. This phase is only partly distinguishable (?) in North America, but very well marked in the Atlantic, where polar water masses (perhaps with seasonal drift-ice) re-advanced from Polar Circle to Ireland and even into the Biscaya (Ruddiman-McIntyre, Duplessy 1981) over a distance of more than 2000 km. One of the consequences was a new glaciation of the Scottish highlands, while in southern Germany forests were once more replaced by tundra.

Many new data (Eicher-Siegenthaler, Duplessy 1981) indicate a quite abrupt beginning of the warm Alleröd as well as of the cold Younger Dryas. A particularly convincing evidence gave investigations on the change of beetle faunas and their climatic implications : changes of summer temperatures of the order of 10-12° C in less than 100 years seem to be reliable (Coope 1977). One of the possible interpretations (Grosswald, Denton-Hughes) is the sudden disintegration of large ice-masses from marine -based (and thus conditionally unstable) ice-sheets, perhaps in the Barents-Sea or the Kara-Sea. As giant table ice-bergs (with a total volume of several $10^6 km^3$), they floated into the North Atlantic, probably responsible for the strong spreading of polar water masses southward, in remarkable contrast to the radiation optimum as an orbital effect. The warming at the beginning of Preboreal (about 9.8 ka ago) was also rather abrupt (Duplessy, Eicher-Siegenthaler).

Simultaneous with the Bölling-Alleröd warm phase, a sudden moist phase has been observed in many subtropical high mountains like Mexico (Heine, with catastrophic rainfall) and Hindukush/northwestern Himalaya, initiating a short and intense peak of glaciation. Combining this evidence with the information from marine cores in the Indian Ocean and the vegetation history of NW-India and Pakistan, the large-sacle Asian-African Monsoon system reveals a surprising evolution (cfr next chapter). During this warm phase 12-10 ka the monsoon circulation intensified to a degree higher than today : stronger ocean upwelling along the Arabian coast (Prell), higher runoff into the Bay of Bengal (Duplessy 1982a, b), large lakes all over North Africa (Street-Perrott, Rognon, Maley 1980b), in Arabia and NW Italia (Singh). This moist period agrees well with Kutzbach's model; at least in Africa it lasted (with a marked interruption about 7500 BP); well into the mid-Holocene, while the final desiccation of the Sahara occurred about 5.5 ka, of Rajasthan about 4 ka BP. In its main phare around 10 ka, the seasonal drift-ice of the Subantarctic was reduced by about 600 kms in comparison to now : this weakened significantly the southern hemisphere circulation, while the northern circulation, at least in the Atlantic section, may have been stronger than now, but weaker than during the glacial.

Consequence of these circulation changes for the air-sea interaction along the equator and the composition of the atmosphere will be considered in a later chapter. For our purpose the following facts are remarkable :

4) A marked moist phase in the Asian-African monsoon region, with an intensification and expansion of the monsoon system between 12 and about 8 ka;
a coincidence with the Lake Bonneville-Lahontan Phase in western USA and
with the glaciation of subtropical high mountains remains to be verified.

5) In the Atlantic-European section the warm Alleröd-Bölling phase lasts about
2000 years; its beginning and its end are apparently quite abrupt (around
100 years or less), the latter with a marked cold phase around 10.5 ka
(see alson Appendix).

Last Glacial (25-14 ka) : Building and Decay

At the present level of knowledge, the total number of glacials and interglacials is rather uncertain : while some ocean cores or loess profiles
suggest a number near 20 (or more, dependent on definition), the geological
evidence in the vicinity of ice-sheets and glaciated mountains seems to
indicate not more than four major glaciations (? cfr Schäfer). This discrepancy (Kukla) is as yeat unsolved. From the point of view of a climato-
logist, however, these questions are less important than the geophysical
problem : under what conditions can large ice-sheets grow and decay ? This
can be studied preferably on the evidence from the last glacial, which is
more complete than from all earlier glacials together.

The accumulation of a continental ice-sheet necessarily starts with an extended snow-cover surviving one melting season; after such an event the
probability of summer melting will decrease asymptotically, with increasing
snow depth, towards zero. The well-known ice/snow albedo-temperature feedback (W. Kellogg) intensifies this process over synoptic-scale areas (10^5-
10^6 kms), while over smaller areas (10^2-10^4km^2) it could be less effective.
The increase of the plateau ice at Baffin Island from 37 000 km^2 to about
140 000 km^2 during the three centuries of the Little Ice Age was obviously
insufficient to start an irreversible process of glaciation. With an annual
precipitation of 30 cm (water equivalent) the building of a 1000 m ice-sheet
latst about 3000 years - but its climatic effect (radiation, heat conduction) -
depends more on its area than on its thickness. The accumulation of the last
glacial started in two rather rapid steps around 115 ka (Stage 5e/5d) and
75 ka (Stage 5a/4)(Ruddiman et al. 1980) : the continental ice growth curve
derived from benthonic micro-fossils leads the SST-curve derived from
planktonic micro-fossils by some 4 ka, while the SST cold peak lags less
than 2-3 ka after the ice-volume maximum of Stage 4. At the western North
Atlantic SST dropped just after this maximum by about 9° C. During the ice
growth phase the Atlantic SST was only 1-2° C cooler, near Ireland perhaps
even warmer than now, with the same salinity; there is no indication for a
cold Gulf Stream. The inflow of relatively warm moist air was most instru-
mental in the accretion phase of the ice-sheet; during this period a very
intense baroclinic zone must have persisted (1.c. Fig. 8) along Labrador/
Newfoundland and the East Greenland Current, with frequent cyclones turning
to NW (Hudson Bay area) or NE (Scandinavia). This situation should have
prevailed during the whole year, while seasonal sea-ice covered large portions of the Norvegian Sea (T.B. Kellogg 1978) : together with frequent
blocking anticyclones at the British Isles this should favour an anti-
cyclonic track of cyclones towards SE into East-Central Europe. The final
ice growth phase after 25 ka contributed hardly more than about 10 percent
of the global ice volume (Ruddiman 1980 Fig. 1); the local distribution
may have been different from the earlier ice growth phases.

At the peak of the last glaciation (about 18 ka) a marked transition from
an ice-growth mode to an ice-decay mode must have occurred, while the above-
mentioned positive feedback process became ineffective, contrary to expec-
tation. What processes are responsabile for such a strange shift in the
mass-budget ?

This enigma has now found a rather convincing solution (Ruddiman and McIntyre
1981). During the maximum extent of ice-sheets (possibly including parts of
the shallow Barents and Kara shelf (Grosswald) meltwater and calving icebergs
reached the northern Atlantic, cooling it substantially during and after the
glacial maximum. Now the sea-ice area increased strongly : polar waters with
summer SST of 2-4° C and seasonal sea-ice extended - until about 13 ka BP -
over the whole North Atlantic up to a line from Portugal to Long Island,
most probably along about Lat. 44° N including the Bay of Biscay. This
happened together with a marked drop of surface salinity, increasing vertical stability, reducing biological productivity (carbonates) and evaporation. Now the continental ice-sheets were undernourished and gradually
shrunk, retaining their original area and their climatic role during the
period 16-13 ka. In contrast to this, the Pacific was neither reached by
icebergs nor by much glacial wastage; its climate remained more genial.

One of the important consequences of this advance of polar water masses to
Lat. 43-45° N, together with the extension of the Laurentide ice-sheet to
Lat. 40° N, was the increase of tropospheric baroclinicity, the intensifica-
tion of the atmospheric circulation and of the wind-driven ocean circulation.
Frequent and intense cyclones were travelling into the Mediterranean.
Increasing wind stress (Sarnthein-Tetzlaff 1981) intensified oceanic upwelling
along the equator and along those coastlines, where the winds were running
parallel to the coast, with lower pressure above the heated continents.
With increasing Ekman pumping, equatorial and coastal upwelling intensified
together. Since the Subantarctic drift-ice simultaneously advanced some 500-
600 km equatorwards of its present position (Hays et al. 1978), the southern
hemisphere circulation was also intensified.

As a further consequence, the water vapour content of air was reduced;
nearly everywhere, especially in low latitudes, evidence of more arid climates
exists (Sarnthein), and the humid tropical rain-forests retreated to a few
refuges. Due to upwelling in equatorial Pacific and Atlantic, to the general
lowering of SST, the sinking sea-level and the increase of polar sea-ice,
global evaporation (and precipitation) were reduced by some 20-25 percent
(Flohn 1983). The continental surface temperatures were lower than now,
i.e. fluxes of latent and sensible heat both decreased. In addition to the
increase of surface albedo (from about 0.14 to 0.20) the remaining surplus
of solar radiation has been used to warm the upwelling ocean water.

At present, upwelling cool water drastically reduces evaporation of tropical
oceans - from 4-5 mm/d to 0.8-2 mm/d (Henning-Flohn). Furthermore, its high
biological productivity consumes much CO_2 : in five years with low SST the
annual growth rate of atmospheric CO_2 is just one half of that in five
other years with high SST (Flohn 1981, 1982; cfr Bacastow et al., Angell,
W.C. Clark). This coincides well with the measurements of fossil CO_2 from
Greenland (Neftel et al.) and Antarctic ice cores (Delmas et al.) : during
(and after) the 18 ka-peak the global CO_2 content was not higher than about
200 ppm, in contrast to about 295 ppm around 1900 and 340 ppm now. (See also
Dansgaard-Oeschger, this volume, and Appendix).

During this cold-arid phase (18-13 ka) the available evidence indicates that
the Asian-African monsoon system was weaker than today : decreasing runoff

into the Gulf on Bengal, less intense upwelling along the Arabian coast
(Prell, Duplessy 1981a, b). While recent results (Hövermann-Kuhle) indicate
a strong glaciation of northeastern Tibet, glaciation of Himalaya and
Karakorum was apparently limited to the mountains itself, while the central
and western part of the Tibetan plateau acted, during this arid phase, still
as a summer-time heat source.

From these reasons the following facts are of particular interest :

6) During the growth phases of the last glacial (especially about 75 and 25-
   20 ka BP) the Atlantic remained nearly as warm as now.

7) The glacial maximum (20-18 ka) with its trend reversal from ice growth
   to ice decay coincides with an increasing flux of glacial meltwater,
   drift-ice and icebergs into the Atlantic up to about Lat. 44° N, reducing
   evaporation, ice-sheet nourishment and cyclone tracks.

8) This situation leads to increasing circulation, concentrated in middle
   and low latitudes, to predominant upwelling, reduction of the CO_2 and
   H_2O content of the atmosphere, a world-wide arid phase (until about 13 ka)
   and an alteration of the radiation and heat balance.

9) During this arid phase the glaciation of the Tibetain plateau intensified
   only about the northern rim : the Asian-African monsoon system was weakened
   but maintained in substance.

The last Interglacial (130-120 ka BP)

During the last interglacial – defined as Eem sensu stricto or Stage 5e
Emiliani-Shackleton – sufficient evidence is available that it was in several
areas slightly warmer and more humid than in the Early Holocene. Temperatures were 2-2.5° C higher than today; a remarkable evidence of a warmer
climate in Western Europe were the bones and teeth of hippos in the fluvial
deposits of the Thames and of the Rhine (Koenigsberger), probably indicating
a genial winter climate without longer (or severe) frost periods.

One of the most noteworthy facts was the occurrence of a sealevel about
5-7 m higher than now, convincingly evidence at many far-distant sites
(Hollin). This apparently happened right in the middle of this 10 ka period,
accompagnied by a local cooling of SST in the southern Indian Ocean, but un-
related to the rose of the global ice volume starting several (perhaps 5-6)
ka later (Dansgaard-Duplessy, Aharon et al.). It seems to be generally
agreed that this sea-level rise was initiated by a disintegration of the
marine-based part of the West-Antarctic Ice during the peak of the last
interglacial. However, its time-scale and local mechanism (a calving bay as
suggested be Denton-Hughes ?) is still controversial and shall not be discussed here.

In the European area this glacio-eustatic sea-level rise resulted in large
changes of the land-seaèdistribution, particularly between the Baltic and the
White Sea, where a shallow sea isolated Scandinavia and Finland from the
continent. A large incursion of the sea occurred also along the rivers Ob
and Yenisey, up to Lat. 61° N. Both features lead to an amelioration of the
harsh climate of northern Russia and western Siberia (Frenzel); precipitation were almost everywhere higher than now, and the permafrost boundary
retreated several 100 kms farther to the north than in the Holocene.

Unfortunately relatively little information is available from North America and other continents; SST data from ocean cores are only rarely significantly warmer than during the Holocene.

The essential fact can be summarized as follows :

10) The warmest part of the last interglacial may have been slightly warm than the Holocene Optimum. Probably caused by a collapse of the marine-based part of the West Antarctic ice-sheet, the global sea-level rose in the middle of this period by 5-7 m, isolating Fennoscandia and penetrating deeply into Western Siberia.

Late Cenozoic (about 14-3.5 Ma) : ice-free Arctic versus glaciated Antarctic

One of the most remarkable discoveries of the Deep-Sea Drilling Program was the early onset of local Antarctic glaciations, at or before the Eocene-Oligocene boundary (38 Ma) (Kennett 1977). Simulatneous with a rather gene-ral cooling of the tropical oceans during the mid-Miocene (16-13 Ma ago : Frakes-Shackleton 1980) the East Antarctic continent glaciated completely, while about 8-9 Ma ago this ice-sheet sread to the West Antarctic archipelago and the Ross Sea shelf. The Antarctic ice reached its greatest volume about 6-5 Ma ago, with characteristic fluctuations at orbital time-scales[*].

During this time, and probably also later until mid-Pliocene (about 3-5 Ma ago) the Arctic Ocean was essentially ice-free. A few scattered erratica (D.L. Clark) may indicate some marginal glaciations, such as during Miocene (9 Ma) in Alaska. But the occurrence of mixed boreal forests in Ne-Siberia, at the New Siberian Islands and Banks Island (Frenzel, Hopkins, Wolfe) indicate, along the coasts, summer temperatures near 12° C which are incon-sistent with a permanent ice-cover. 5 Ma ago lived, at the Alaska coast (Lat. 66° N), in a rich coniferous forest, the same insect fauna which is now characteristic for coastal British-Columbia (48-54° N). Evaluation of a long core in the equatorial Pacific (Shackleton-Opdyke) indicated that fluctuations of the global ice volume increased gradually since about 3.6 Ma. This is interpreted as the beginning of the Northern Hemisphere glaciation, with volume maxima of about 20-25 percent of the last glacial.

This evidence convinces us that during a period of 10-12 Ma, a fully glaciated East Antarctica coexisted simultaneously with an ice-free Arctic Ocean. The glaciation of the northern continents - with the first nuclei probably at Greenland, Baffinland and Iceland (here confirmed about 2.5 Ma ago) (perhaps also at what is now the Barents Sea Shelf) - occurred at the timen when tectonic uplift together with volcanic activity gradually closed the Panama Isthmus. This was completed about 3.5 Ma ago, when a separate evolution of oceanic life in both Atlantic and Pacific began, while the land vertebrates of North and South America migrated freely across the bridge. This closure blocked the Northern Equatorial Current with its tropical water, deviating it towards north, thus strengthening the Gulf Stream and the transport of atmospheric moisture towards NE. This intensified cyclogenesis around Newfoundland, especially during winter; precipitation in the Greenland-Labrador area should have increased substantially. This can be considered as a trigger initiating the northern hemisphere glaciation. Repeated inflow of glacial meltwater were one of the essential prerequisites of a stratified Arctic Ocean with a shallow low-saline deck layer, leading first to a seasonal and finally to a permanent sea-ice cover. Actually, the Arctic Ocean is

* Footnote see p. 25

characterized by a 70 percent permanent ice-cover, while in the well-mixed Subantarctic Ocean this is restricted to 15-20 percent, mainly in the Weddell Sea area.

Wat climate patterns can respond with such a strange unipolar glaciation ? First we should remember that the present earth also has an asymmetric climate. In contrast to the present geometry of solar radiation (with a surplus of solar radiation above the southern hemisphere during its summer) the troposphere above Antarctica is annually about 11° C colder than above the Arctic (Flohn 1967, 1978) (Fig. 2).

This is a consequence of the different heat budget of an isolated Antarctic and of an Arctic ocean with its thin ice-cover and its large heat capacity (Flohn 1978). While above the Subarctic quasi-stationary eddies are frequent, caused by the existing two continents and two oceans, the meridional heat transport is here twice as large as above the Subantarctic, which is also hampered by the lack of oceanic heat advection, high surface albedo and low cloudiness.

At a rotating sphere with a uniform atmospheric envelope, the intensity of the circulation depends on V. Bjerknes law (1897), combining the vertical and the horizontal gradients e.g. of temperature and pressure. The vertical tropospheric lapse rate is quasi-stationary, depending on the radiative properties of such gases as H_2O, CO_2 and O_3. Since the heating difference equator-pole is asymmetric, so is the circulation today : except during northern winter, the southern circulation is stronger and crosses the equator into the northern hemisphere, with an annually averaged position of the "meteorological equator" at Lat. 6° N (Flohn 1978, Ramage 1982). Due to the variation of the Coriolis parameter $f = 2 \Omega \sin \varphi$ with latitude φ, we observe a number of anomalies in the vicinity of the equator caused by this asymmetry including the patterns of oceanic upwelling caused by Ekman "pumping" and the strong baroclinic counter-current at Lat. 4-8° N. All climatic zones and wind regimes are situated farther equatorwards at the souhtern hemisphere. The position of the subtropical jet and the subtropical anticyclones - as tje boundary between the extratropical anticyclones - as the boundary between the extratropical circulation and the tropical Hadley cell - is controlled by the isobaric difference equator-pole (see Fig. 3).

If the Arctic sea-ice should disappear, this asymmetry must increase even more. A quantitative assessment of the shift of the subtropical belt (and other climatic boundaries) on an empirical base (Flohn 1982) is hampered by some complexities. Taking into account the present temperature distribution above a frozen Arctic and that expected above an ice-free Arctic, one should assume that the decrease of the temperature gradient and the shift of climatic belts is much stronger during winter (with a rise of surface temperatures from -32° C to about 0° C) than during summer (with a rise from 0° C to perhaps 8-10° C). The poleward displacement of the norhtern subtropical arid belt is estimated to be 5-6° C Lat. in winter and 2-3° C Lat. in summer : this leads to a reduction of the subtropical winter rains, notably in the Mediterranean, Near and Middle East and in California. Similarly the Intertropical Convergance Zone should have shifted further north, towards Lat. 9-10° N.

Estimates of the climatic change to be expected in such a case have been made in different forms :

a) evaluation of the palaeoclimatic data from the Pliocene, as given by a
   working group from the USSR under Budyko;

b) simulation with different models (e.g. GFDL : Manabe-Stouffer, Manabe-
   Wetherald, Wetherald-Manabe, Manabe-Wetherald-Stouffer 1981), assuming
   quadrupling of CO_2. None of these models takes into account oceanic
   circulation and heat transport; such coupled models are in development,
   but no systematic results are as yet available. A seasonal disappearance
   of Arctic sea-ice results in Manabe-Stouffer (cfr als Parkinson-Kellogg),
   but the prescribed depth of the mixed-layer ocean prohibits a transition
   to a permanently open state.

c) semi-empirical extrapolation of geophysical relationships, combined with
   some model results (Flohn in W.C. Clark).

There is some coincidence between these different estimates, especially
towards increasing aridity near Lat. 40° N (Fig. 4). The author assumes an
increasing role of the southern hemisphere circulation crossing the equator,
with an intensification and/or prolongment of oceanic upwelling just south
of the equator. However, several important changes of boundary conditions
prohibit the direct use of palaeoclimatic data for estimating a possible
future evolution of our climate. 3-5 Ma ago most mountains existed only in
a rudimentary stage; this is especially true for the Alps, here evidenced
by a relatively smooth relief now in altitudes between 2000 and 4000 metres.
An important fact is the late uplift of the Tibetan Plateau, starting
1.8-2.4 Ma ago, while during mid-Pliocene the altitudes had been hardly
higher than 1000 m. At this time, the present monsoon system cannot have
existed (Hahn-Manabe 1975). Its evolution seems to coincide with the step-
wise desiccation of the formerly humid Sahara (McCauley) and the southward
shift of African tropical rain-forests to both sides of the equator (Maley
1980).

The essential palaeoclimatic facts may be summarized as follows :

11) During the late Cenozoic (about 14-3.5 Ma ago) Eastern Antarctica was
    completely ice-covered; this ice-sheet gradually extended over Western
    Antarctica and the Ross Sea Shelf. At this time the Arctic Ocean was
    essentially ice-free.

12) This unipolar glaciation caused an asymmetric climatic zonation, with a
    northward shift of all northern hemisphere climatic belts. Most important
    was the migration of the subtropical arid celles, causing a reduction of
    the subtropical winter rain belts.

13) The continental glaciations at the northern hemisphere developed after
    the closure of the Panama Isthmus strengthening the Gulf Stream system.
    As a consequence, the Arctic Ocean became gradually stratified with a
    low-density meltwater-layer, finally causing the formation of a
    permanent ice-cover.

* A more detailed chronology of the palaeoclimatic evolution at the upwelling
  region off NW-Africa - distinguishing 11 climatic steps during the last 24
  MA (earliest Miocene) - is given by M. Sarnthein et al. in : U. von Rad et
  al. (Eds.) : Geology of Northwest African Continental Margins.
  Heidelberg 1982 (Springer) pp. 584-604, cfr also D. Schnitker : Earth
  Science Rev. 16 (1980), 1-20, Nature 284 (1980), 615-616.

Appendix : Abrupt climatic changes

From the existing literature no clear definition of "abrupt changes" could
be found. As a working definition the following can be proposed : an abrupt
climatic change has a time-scale in the order between 50 and about 200 years,
while the temperature difference is in the order of half of the difference
between glacial and interglacial, i.e. 2-3 K.

Some examples from the transition Late-Glacial -Early Holocene (ca. 13-7 ka
BP) have been discussed :

a) Warming at the beginning of the Bölling/Alleröd warm phase (ca. 13 ka)
   probably identical with termination IB (Duplessy), with the onset of a
   humid phase in northern Africa and other areas in Lat. 10-25° N including
   maximum extension of mountain glaciers in this belt. Freshwater inflow
   to the ocean leads to increased evaporation, but to a reduction of vertical mixing.

b) Cooling at the transition between Alleröd and Younger Dryas (ca. 11 ka)
   with marked readvance of polar water in the Atlantic and of glaciers in
   the Scottish highlands. This event happened simultaneously with the
   shortest distance sun-earth (perihelion) during northern summer; the
   hypothesis of a glacial surge in the Barents Sea/Kara Sea area with out-
   flow of icebergs and glacial meltwater should be considered.

c) Warming between Younger Dryas and Preboreal (ca. 10 ka), probably inden-
   tical with termination IA (Duplessy). In both warming episodes the CO_2
   content at Dye 3 (S. Greenland) increased suddenly about 40 ppm
   (Dansgaard-Oeschger).

d) Simultaneous short dry episodes over northern Africa from Mauretania to
   Ethiopia (Rognon) around 7.5 ka, simulaneous with a glacial surge in the
   Hudson Bay area with a sea-level rise of 7-10 m at the European coasts;
   no evidence for ocean cooling (Canary current !) is available.

Since the two warming events are correlated with an abrupt increase of
atmospheric CO_2, the hypothesis of a feedback effect : warming - wealer
Hadley circulation - prevailing downwelling and rapid increase of CO_2 and
H_2O from equatorial oceans - further warming (Flohn 1981, 1982) should be
investigated. A reversed feedback mechanism would occur after a glacial
iceberg-meltwater surge causing cooling and stronger circulation - prevailing
upwelling (decreased atmospheric CO_2 and H_2O. Saltzman et al. developed
recently a simple coupled climate model, which produced - after adding
stochastic perturbations - a bimodal, almost intransitive auto-oscillatory
type of climate which could put these positive feedbacks into a more general
frame.

These events can serve as examples for further research. Of primary interest
are quantitative estimates of the climatic changes involved; errors arising
from the varying migration speed of trees should be taken into account.
Furthermore, the chronology of the events (which could partly be time-
transgressive, e.g. the onset of the Younger Dryas) should be refined; marine
and terrestrial events should be carefully compared with the aim of a
coherent sequence of climatic episodes.

Selected References

P. Aharon et al. : Nature 282 (1980), pp. 649-651.

J.K. Angell : Mon. Weather Review 109 (1981), pp. 230-243.

R.B. Bacastow et al. : Science 210 (1980), pp. 66-68.

A.L. Berger : Quatern. Res. 9 (1978), pp. 139-167; Vistas in Astronomy 24
        (1980, pp. 103-122.

M.I. Budyko (unpublished report, 1981).

D.L. Clark et al. : Spec. Paper Geol. Soc. America 181 (1980).

W.C. Clark : Carbon Dioxide Review 1982. Oxford 1982 (Clarendon Press).

G.R. Coope : Philos. Transact. Roy. Soc. London B 280 (1977), pp. 313-340.

W. Dansgaard, J. Cl. Duplessy in : H. Flohn, R. Fantechi (Eds.) : Whither
        Climate ? (in print).

A.F.M. De Jong, W.G. Mook, B. Becker : Nature 280 (1979), pp. 48-49.

R.J. Delmas, J.M. Ascencio, M. Legrand : Nature 284 (1980), pp. 155-157.

R.H. Denton, T.J. Hughes (Eds.) : The Last Great Ice Shields. New York 1981
        (Wiley-Interscience).

J. Cl. Duplessy et al. : Palaeo^3 35 (1981) : pp. 121-144; Nature 295 (1982a),
        pp. 494-498; 296 (1982b), pp. 56-59.

U. Eicher, U. Siegenthaler : Boreas 5 (1976), pp. 109-117; Quatern. Res.
        15 (1981), pp. 160-170.

H. Flohn in : E.M. van Zinderen Bakker (Ed.) : Antarctic Glacial History and
        World Palaeoenvironments. Dordrecht 1978 (Reidel), pp. 3-13; Ann.
        Meteor. 3 (1967), pp. 76-80; Physik. Blätter 37 (1981), pp. 184-190;
        Journ. Meteor. Soc. Japan 60 -1982), pp. 268-273; also in W.C. Clark
        (1982), pp. 143-185.

L.A. Frakes : Climates Throughout Geologic Times. Amsterdam 1979 (Elsevier).

B. Frenzel : Science 161 (1968), pp. 637-649.

M. Gamper, J. Suter, H. Holzhauser : Geographica Helvetica 37 (1982), pp.
        105-126.

M.G. Grosswald : Quatern. Research 13 (1980), pp. 1-32.

D.G. Hahn, S. Manabe : Journ. Atmos. Science 32 (1975), pp. 1515-1541.

J.D. Hays in : E.M. van Zinderen Bakker (Ed.) : Antarctic Glacial History
        and World Palaeoenvironments. Dordrecht 1978 (Reidel), pp. 57-71.

K. Heine : Erdkunde 28 (1974), pp. 303-312; 31 (1977), pp. 161-178; Ibero-
        Amerik. Archiv N.F. 7 (1981), pp. 69-76.

D. Henning, H. Flohn : Contrib. Atmos. Phys. 53 (1981), pp. 423-431.

J.T. Hollin : Nature 283 (1980), pp. 629-633

D.M. Hopkins et al. : Palaeo^3 9 (1971), pp. 211-231.

J. Hövermann, M. Kuhle (in print).

T.B. Kellogg et al. : Boreas 7 (1978), pp. 61-73.

W.W. Kellogg : Journ. Geophys. Res. (in print).

J.P. Kennett : Journ. Geophys. Res. 82 (1977), pp. 3843-3860; Palaeo[3] 31 (1980), pp. 123-152.

J. Kington : Weather 30 (1975), pp. 109-114.

W. Koenigsberger (unpublished lecture).

H. Cl. Korff, H. Flohn : Ann. Meteor. 4 (1969), pp. 163-164.

G. Kukla : Earth-Science Review 13 (1977), pp. 307-374; in A. Berger (Ed.) Climatic Variations and Facts. Dordrecht 1980 (Reidel), pp. 207-232.

J.E. Kutzbach : Science 214 (1981), pp. 59-61; Journ. Atmos. Sci. 39 (1982), pp. 1177-1188.

H.H. Lamb : Climate, Past, Present and Future. Vol. 2, London 1977 (Methuen); Climate History and the Modern World, London 1982 (Methuen).

H.H. Lamb : Quatern. Res. 11 (1979), pp. 1-20.

J. Maley in : M.A.J. Williams, H. Faure (Eds.) : The Sahara and the Nile. Rotterdam 1980a, pp. 63-86; Thèse Univ. Montpellier, 1980b.

S. Manabe, R.J. Stouffer : Journ. Geophys. Res? 85 (1980), pp. 5529-5552; Nature 282 (1979), pp. 491-493.

S. Manabe, R.T. Wetherald : Journ. Atmos. Sci. 37 (1980), pp. 99-118.

J.F. McCauley et al. : Science 218 (1982), pp. 1004-1020.

D.A. Mooley, G.B. Pant in : T.M.L. Wigley et al. (Ed.) : Climate and History. Cambridge 1981 (Univ. Press), pp. 465-478.

R. Neftel et la. : Nature 295 (1982), pp. 220-223.

Sh. Nicholson in : T.M.L. Wigley et al. (Ed.) : Climate and History. Cambridge 1981 (Univ. Press), pp. 249-270.

C.E. Parkinson, W.W. Kellogg : Climatic Change 2 (1979), pp. 149-162.

Chr. Pfister : Das Klima der Schweiz von 1525-1860 und seine Bedeutung in der Geschichte von Bevölkerung und Landwirtschaft (1983, in print).

W.L. Prell et al. : Quatern. Research 14 (1980), pp. 309-336.

C.S. Ramage et al. : Journ. Geophys. Res. 86 (1981), pp. 6580-6598.

P. Rognon : Palaeoecology of Africa 13 (1981), pp. 21-44.

W.F. Ruddiman, A. McIntyre : Science 212 (1981), pp. 617-627; Palaeo[3] 35 (1981) pp. 145-214; Quatern . Res. 16 (1981), pp. 125-134; Climatic Change 3 (1980), pp. 65-87.

B. Saltzman et al. : Tellus 30 (1980), pp. 572-584; 34 (1982), pp. 97-112; Journ. Atmos. Sci. 38 (1981), pp. 494-503.

M. Sarnthein et al. : Nature 271 (1978), pp. 43-46; Nature 293 (1981), pp. 193-196.

M.E. Schlesinger : OSU Climate Research Paper, n° 40 (1982).

I. Schäfer : Eiszeitalter und Gegenwart 32 (1982), pp. 213-216.

N.J. Shackleton, N.D. Opdyke : Nature 270 (1977), pp. 216-219.

N.J. Shackleton in A. Berger (Ed.) : Climatic Variations and Facts, Dordrecht (Reidel) 1980, pp. 167-179.

D.G. Singh : Philos. Transact. Roy. Soc. London B 267 (1974), pp. 467-501.

F.A. Street-Perrott, A.T. Grove : Quatern. Res. 12 (1979), pp. 83-118.

R.T. Wetherald, S. Manabe : Journ. Geophys. Res. 86 (1981), pp. 1194-1204;
    Climatic Change 3 (1981), pp. 347-386 (with Stouffer).

J.A. Wolfe : Palaeo^3 30 (1980), pp. 313-323.

Palaeo^3 = Palaeogeography, Palaeoclimatology, Palaeoecology

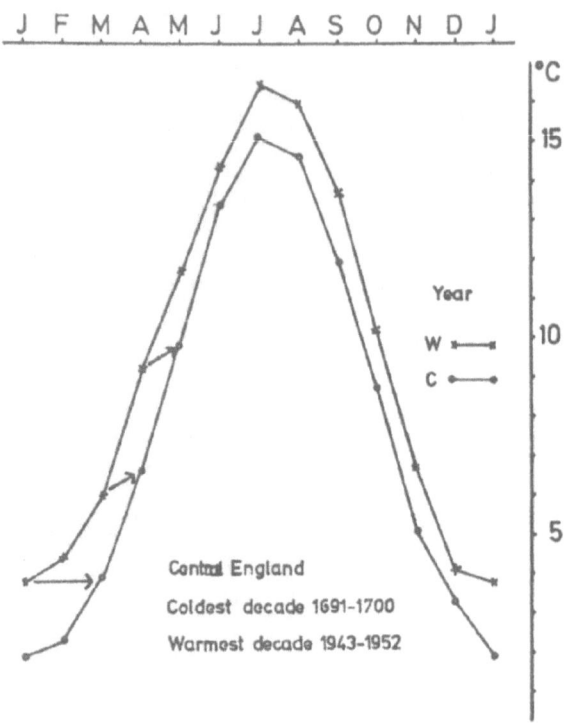

J F M A M J J A S O N D J

°C

15

Year

W ×——×   10

C ●——●

5

Central England
Coldest decade 1691-1700
Warmest decade 1943-1952

Fig. 1 : Central England Temperature Record (Manley 1973); decadal averages
for the coldest and the warmest decade of the record. Note time
difference of 1-2 months during late winter and spring temperature.

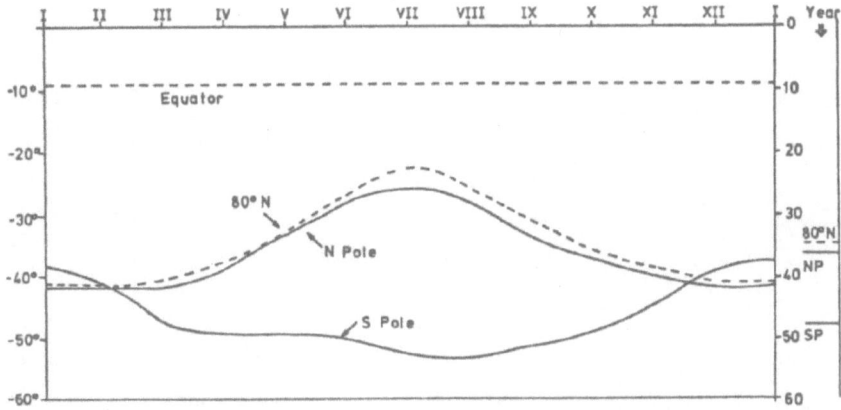

Fig. 2 : Average seasonal temperature course of the 300/700 mb layer : 7 stations along the Equator (3° S - 7° N), 7 stations along Lat. 80° N,
both averaged for 6–10 years each, North Pole (T 3, difference to
the floating ice stations from the USSR negligible) and Amundsen-
Scott station (surface pressure 681 mb, here taken as 700 mb),
period generally 1957–66 (Flohn 1967). Note the near-equal values
at both poles during northern winter/southern summer.

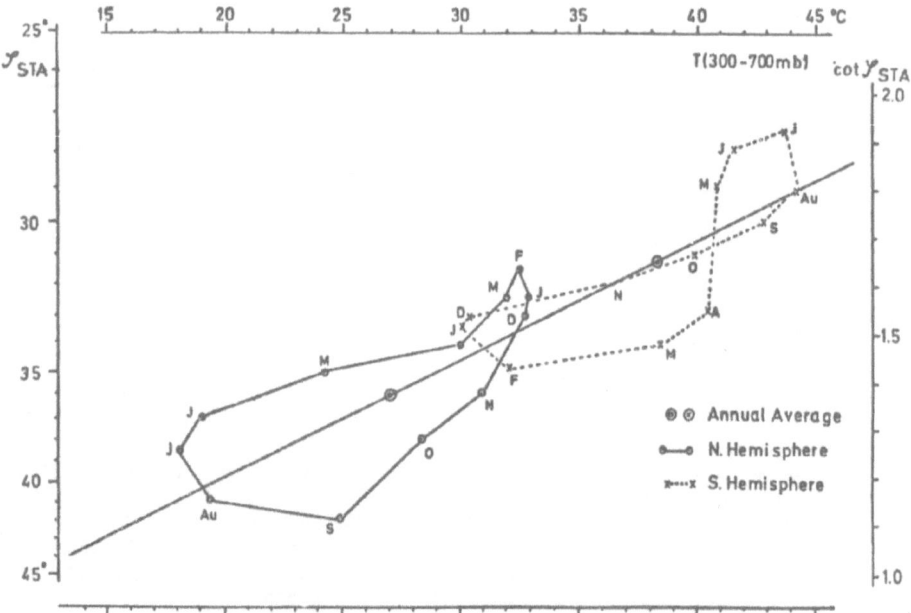

Fig. 3 : Tropospheric (300/700 mb layer) temperature difference Equator-Pole
        for each month at both hemispheres and simultaneous position (φ_{STA})
        of the subtropical anticyclonic belt (linear scale of cot φ_{STA})
        (Korff-Flohn 1969).
        Correlation coefficient between both independent sets + 0.85, with
        pressure index lagging 1-2 months around + 0.91.

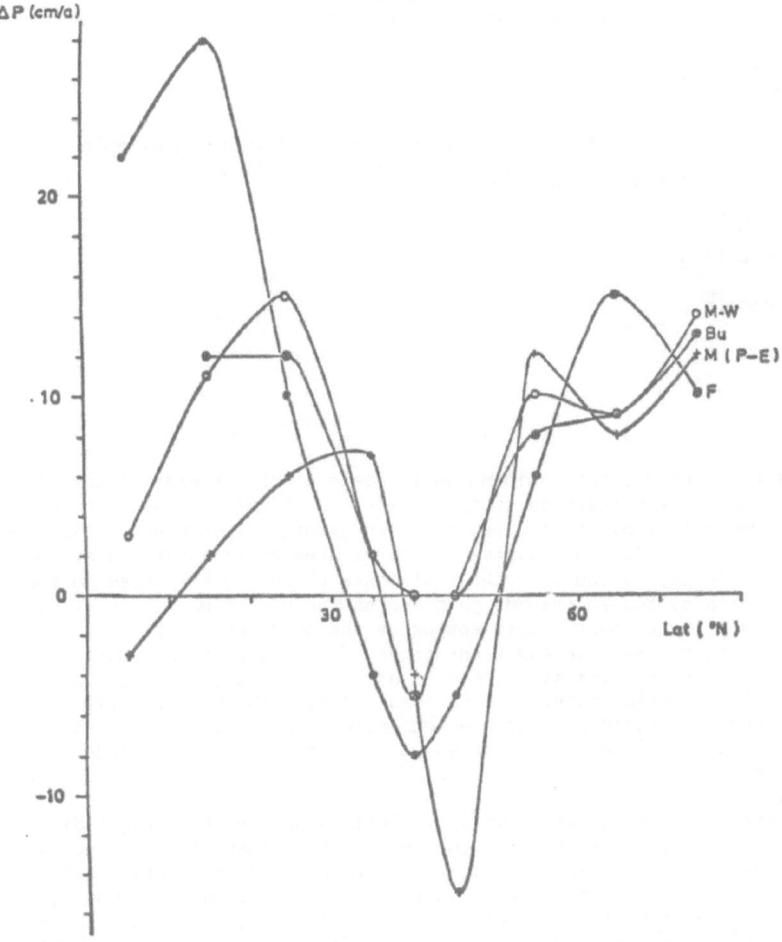

Fig. 4 : Desiccation around Lat. 40° N, comparison of results.
Change of average precipitation (relative to recent values) according
to palaeoclimatic data from Pliocene (Budyko 1981), model results
from Manabe-Wetherald (1980, 4 x CO_2, but no ice-free Arctic;
P-E = precipitation - evaporation = runoff) and semi-empirical
estimates (Flohn 1981, publ. 1982).

THE PHYSICAL BASIS OF CLIMATE MODELLING

by J.F.B. MITCHELL
UK METEOROLOGICAL OFFICE

1. Introduction

2. Physical basis of modelling, using atmospheric models as an example

   (a) Dynamics and physical parametrization

   (b) Using a model

   (c) Verification

3. Climate models

4. Summary

1. Introduction

Over the past decade, three-dimensional models of the general circulation have been used extensively to study climate and climate change. The use of such models is limited to a few research groups. This paper is intended to clarify the physical principles which are used to formulate numerical models of climate, to indicate the confidence which can be placed in the parametrizations, and to provide guidance on the use of models in climate studies. A climate model should represent the interaction of the atmosphere, ocean, sea-ice and land-ice and the land surface. Until recently, most three-dimensional models used for climate studies consisted of general circulation models of the atmosphere, with an interactive land surface, and prescribed sea surface temperatures and sea ice extents. Hence, this paper is mainly concerned with general circulation models of the atmosphere.

The climate at a particular location is defined by the statistical distribution of temperature, rainfall, wind and so on at that place. Thus, we are not only interested in the mean temperature for a particular month of the year, but how much temperature is likely to vary within the month, and how much the monthly mean temperature is likely to vary from year to year. Observational studies have used periods of order 30 years to define surface climate statistics, (for example see Schutz and Gates, 1971), but more detailed studies, particularly those using upper air data (for example Oort and Rasmusson, 1971) often use a shorter period. Because of the expense of running computer models, statistics from numerical simulations are gathered over only a few years. In the examples discussed later only three years of model data have been used (Mitchell, 1983).

The circulation of the atmosphere arises because heating (and cooling) by radiation sets up temperature gradients which would not be in equilibrium with a stationary atmosphere. The radiation balance of the atmosphere is shown in Figure 1. Solar radiation is absorbed by the atmosphere, and reflected by clouds or by the earth's surface; the rest is absorbed by the surface (Fig. 1). This energy is removed from the surface by thermal (long wave) radiation, or as sensible heat (conduction), or latent heat (evaporation). The heat and moisture are redistributed throughout the depth of the atmosphere by both large scale ascent and descent and small scale vertail motions (convection). Both surface and atmosphere emit long

wave radiation to space, so that on average, the net incoming solar
radiation is balanced by the outgoing long wave radiation.

As low latitudes receive more solar radiation than high latitudes, the
tropics are warmer than the poles. If the earth did not rotate about its
polar axis, warm air would rise at the equator, travel towards high
latitudes and descend over the poles. However, the rotation of the earth
produces a much more complicated and variable pattern of motion, or
general circulation.

2. Physical basis of climate modelling, using atmospheric models as an
   example

   (a) Dynamics and physical parametrization

   In a numerical model, the atmosphere is represented by values of wind,
   temperature and humidity which are held at various levels on a mesh
   of grid points which cover the globe. The Meteorological Office
   general circulation model considered here has 5 layers in the vertical
   and 4,626 grid points, each about 330 km apart (for a more detailed
   description of this model, see Corby, Gilchrist, and Rowntree, 1977).
   At any one time, the state of the atmosphere is represented by about
   10^5 numbers. Given the values of these variables at some time, t,
   the values at some future time, $t + \Delta t$, can be obtained using the
   appropriate equations. In atmospheric models, Δt is usually about
   10 minutes, each of the variables being updated about 150 times per
   day. The Meteorological Office 5-layer model takes 15 minutes of CPU
   on an IBM 360/195 for each simulated day.

   The equations used to step forward or integrate the variables in time
   are based on the laws of classical physics. They are derived from the
   Navier Stokes equations (developed from Newton's second law of motion),
   a continuity equation which ensures that mass is conserved, the hydro-
   static equation, the perfect gas law and the second law of thermodynamics.

   In General, the rate of change $\frac{\delta X_i}{\delta t}$ of a variable X_i is a function F_i
   of the variables $\underline{X}$ at and near that location

   $$\frac{\delta X_i}{\delta t} = F_i(\underline{X}, t) \quad X_i = u, v, T, q \text{ at each level} \qquad (1)$$

   where u is the west to east wind, v the south to north wind, T is
   temperature, and q is the humidity.

   Using equation (1), the value of X_i at time t + t may be calculated
   approximately using the value of variables at time t

   $$X_i t(t + \Delta t) = X_i(t) + \Delta t \, F_i(\underline{X}, t) \qquad (2)$$

   Unfortunately, equation (2) cannot be used indefinitely, as it will
   diverge from the correct answer, but the variations of (2) which do
   converge are used in many fields of Fluid Mechanics (see for example
   Richtmyer and Morton (1967) and Haltiner and Williams (1980)).

The term on the right hand side of equation (1) may be expanded

$$\frac{\delta X_i}{\delta t} = \underline{\nabla} \cdot \underline{V} \, X_i + Y + D + N$$

(3)

$$\text{(i)} \qquad \text{(ii)} \qquad \text{(iii)(iv)(v)}$$

Where term

(i)   is the local rate of change of X_i

(ii)   is the change in X_i due to transport by the flow

(iii)   represents the effects of pressure gradients and the earth's rotation in the wind equation, of vertical adiabatic motion in the temperature equation

(iv)   are smoothing (diffusion) terms necessary for numerical solutions of this equation. They may be viewed as representing the effect of small scale motions not resolved on the model grid

(v)   are sources and sinks of momentum (i = u, v), heat (i = T) and moisture (i = q)

Terms (i), (ii) and (where appropriate)(iii) are often referred to as the large scale dynamics. Terms (iv) and (v) represent the effects of processes which cannot be resolved on the model grid and are discussed further below. Most aspects of model simulations improve as the resolution of the grid is increased (there are more grid points) and as more accurate numerical procedures are used in the solution of equation (2). (Manabe et al., 1979).

Most methods for solving (1) require an explicit smoothing term (iv), as otherwise the energy accumulates in the smallest scales resolved by the model grid, and the solutions become unmeteorological. This is usually controlled by some form of diffusion, which may be regarded as a crude parametrization of small scale circulations which cannot be resolved by the model.

The final term in equation (3) covers all the other small scale processes not represented by (iv), and includes boundary layer and surface processes, radiation and convection. For some of these phenomena, for example, radiation, there exists an accepted theoretical treatment which cannot be used with limited computer resources without simplification. There is no accepted theoretical basis for the remainder, so they are represented in simple forms, which embody the main features of the processes concerned.

These parametrizations may be chosen using evidence from one or more of the following sources.

1. Simplified forms of the exact equation (if known).

2. Numerical experiments performed on a finer grid than feasible in the general circulation model.

3. Observations from the atmosphere.

4. Laboratory experiments.

5. Sensitivity integrations made with general circulation model.

The following two examples, concerning the calculation of long wave radiation and cloud amounts, may help to illustrate how the choice is made.

In principle, the flux of long wave radiation at any level z in the atmosphere van be determined exactly from the standard theory of radiative transfer. This explicit integration (Fig. 2, adapted from Rodgers, 1977) includes two integrations in the certical over z, an integration over all angles represented by μ, an integration over all frequencies v, and a summation over all spectral lines i. These can be reduced to a single integration in the vertical, known as the emissivity or grey body approximation, which is used in many general circulation models. Its accuracy may be assessed by compring the flux given by it and the more exact treatments in Figure 2 calculated using the same vertical profile of temperature and absorbing gases. It should be noted that the line strenghts S_i in the explicit integration in Figure 2 are ultimately determinated from laboratory data.

There is no accepted theory for the determination of cloud amount. Hence one must rely on physical intuition to parametrize cloud amounts in the model. Slingo (1980) based her cloud schemes on observations obtained during the GARP Atlantic Tropical Experiment. One would expect that cloud is more likely to form when relative humidity is high rather than low, and this is confirmed by the frequency distribution in Figure 3. Furthermore, the higher the humidity, the greater one might expect cloud cover to be. No simple relationship between relative humidity and cloud amount exists, as illustrated in Figure 4. The frequency distribution of relative humidity in the model may also be different from that of the real atmosphere, so one cannot necessarily base the cloud scheme entirely on observations. Slingo made cloud amount a quadratic function of relative humidity, choosing the adjustable parameter on the basis of both observations and tuning experiments in the atmospheric model.

The sources of the various parametrizations in the UK Meteorological Office 5-layer model are listed in Table 1. Note that apart from the obvious exception of surface reflectivity, each adjustable parameter has the same value over all the 4,626 point covering the globe, so that only a limited amount of tuning is possible.

(b) Using the model

Certain boundary conditions must be specified before the model can be integrated. For an atmospheric model, these will include sea surface temperatures and sea ice extents, orography and the time of year. The usual procedure is then to integrate the model over the period of interest, which may range from a month upwards, and look at the long term statistics of the simulation. In this approach, there are certain assumptions whose validity may affect the relevance to reality of the results from any particular application.

(1) A "mean" state exists in nature and in the model

The evidence presented at this meeting showed that the climate of the earth has varied considerably over a vast range of time scales. Thus, in the real world at least, we do not know if a statistically stationary climatic state exists. However, as stated earlier, climatological data averaged over periods of order 30 years are often used to define a mean climate.

(2) The "mean" state is independant of the initial state

Lorenz (1975) has suggested that the earth's climate may have more than one stable state, and that transitions between these states are rare (climate is almost "intransitive"). It follows that the choice of initial conditions may determine the stable state into which the model first evolves. For example, Watherald and Manabe (1975) found that a simulation made with one of their climate models which commenced with an ice covered surface remained ice covered, whereas when the initial conditions were free of ice, middle and low latitudes remained free of ice throughout the simulation.

(3) The model converges, and converges to the correct solution

Although mathematical models of climate are based on sound physical principles, this alone does not guarantee that the model will produce the desired solution. We cannot check the convergence of such a complex model by carrying out a mathematical stability analysis, so the only way to check if the model converges is to integrate it and look at the results.

(c) Verification

The validity of the model may be assessed by comparing simulations made using present day boundary conditions with climatological data for the appropriate time of year. For example, comparing the mean December to February sea level pressure pattern simulated by the Meteorological Office model (Fig. 5a) with the January climatological pattern (Fig. 5b) one can see that the model reproduces all the main features of the observed circulation. These include the equatorial and mid latitude low pressure belts, and the subtropical anticyclones with form over land in the winter hemisphere, and over the oceans in the summer hemisphre. There are some obvious shortcomings; for example, the northern oceanic depressions are too deep, and pressure over the poles is too high. A more stringent test of the model is to integrate the model through one or more annual cycles to ensure that the para-metrizations are valid at different times of year (see for example, Mitchell (1983)).

One can also compare the variability of real and model data over the appropriate range of time scales. Again, comparing the standard deviation of monthly mean surface pressure from the long term mean from the model (Fig. 6a, averaged over December, January and February) with the standard deviation of the observed January mean (Fig. 6b, northern hemisphere only), one can see that the year to year variability in the model is similar to that in the real world.

The short term variability of the model can also be compared with that of the real atmosphere. For example, Figure 7 shows the power spectral density of daily values of temperature (a) of Central England Temperature for the period 1975-1977, (b) from the lowest layer of the Meteorological Office model at a grid point over eastern England, in each case with the annual and semi-annual cycle removed (Reed, 1983).

(The reader should compare the general shape of the distribution, since some of the peaks may be an artifact of the maximum entropy method used to derive the spectral density). The model produces relatively greater variations with periods of 5 to 10 days than are found in the real data.

Other aspects of the model's performance may be checked by running short forecast integrations from real data. Finally, one can also attempt to reconstruct past climates using the appropriate boundary conditions, although this has the disadvantage that both the boundary conditions and the prevailing climate are not known accurately.

3. Climate models

By using atmospheric models as an example, I have attempted to convey some impression of how climate models are constructed and used. A similar approach is used in modelling the ocean, although there are less data available to guide the development of parametrization schemes, and to verify the model simulations. To be complete, a climate model needs to take into account the interactions between the atmosphere, ocean and cryosphere (see for example Fig. 8). Since the atmosphere, ocean and continental ice sheets respond on time sacles of days, centuries and thousands of years respectively, special care must be taken when representing the interaction between them. Schlesingen (1979) has listed several ways in which an atmospheric model may be coupled to a model of the ocean. Alternatively, Hasselmann (1979) has suggested parametrizing the effect of the quickly changing elements of a climate model (for example, the atmosphere in a model which including a representation of the deep ocean).

4. Summary

The main conclusions reached is this paper are

1. The models discussed above are capable of reproducing the geographical, seasonal and year to year variations in climate.

2. Sub-grid processes may be based on theoretical considerations, observations and laboratory and numerical experiments. Most parameters are chosen on the basis of observations, and there is little scope for tuning.

3. Climate models are expensive in terms of computer time, not only because of their complexity, but also because of their inherent internal variability, which makes it necessary to run long integrations to produce results which are representative.

4. Coupled models of the ocean and atmosphere can be particularly costly to run because of the large disparity in time scales.

Acknowledgement

I am grateful to Dr Peter Rowntree for useful comments on the first draft of this paper

References

Corby G.A., Gilchrist A. and Rowntree, P.R., 1977, United Kingdom Meteorologic-
    al Office 5-level General Circulation Model. Methods in computational
    physics, Vol. 17, pp. 67-110. Academic Press, (New York, San francisco
    and London).

Haltiner G.J. and Williams R.T., 1980, Numerical weather prediction and
    Dynamical Meteorology. (John Wiley and Sons).

Hasselmann K., 1979, On the problem of multiple time scales in climate
    modelling. "Man's impact on climate". Developments in Atmospheric
    Science, Vol. 10 (eds Back, Pankrath and Kellogg), Elsevier, (Amsterdam,
    Oxford, New York), pp. 43-56.

Lorenz, E., 1975, Climate predictability, WMO-GARP Publication Series N^O. 16,
    Geneva, pp. 132-136.

Manabe S., Hahn D.G. and Holloway J.L. Jr, 1979, Climate simulations with
    GFDL spectral models of the atmosphere : Effect of spectral truncation.
    Report of the JOC Study Conference on climate models : Performance,
    intercomparison and sensitivity studies, Vol. 1, GARP, Publication
    series N^O 22, WMO.

Mitchell J.F.B., 1983, The seasonal response of a general circulation model
    to changes in CO_2 and sea temperatures. Quart. J.R. Met. Soc., 109,
    pp. 113-152.

Oort A.H. and Rasmusson E.M., 1971, Atmospheric Circulation Statistics.
    NOAA Professional Paper, US Dept. of Commerce.

Reed D.N., 1983, Model simulation of temperature and precipitation in a
    single grid box in a multi-annual integration of a general circulation
    model. Met O 20 tech. note II/191 (in preparation).

Richtmyer R.D. and MortonK.W., 1967, Difference methods for initial value
    problems. Interscience publishers.

Rodgers C.D., 1977, "Radiative Processes in the Atmosphere". In Proceedings
    of ECMWF Seminars. "Parametrization of Physical Processes in the free
    atmosphere", pp. 5-66.

Schlesingen, M.E., 1979, Comments on ocean-atmosphere coupling and discussion
    of the paper "A global ocean-atmosphere model with seasonal variation :
    Possible application to a study of climate sensitivity". Climate
    Research Institute, Report N^O 39, Oregon State University.

Schutz C. and Gates W.L., 1971, Global climatic data for surface, 800 mb and
    400 mb for January. R 915 ARPA Rand, Corps., Santa Monica.

Slingo J.M., 1980, A cloud parametrization derived from GATE data for use
    with a numerical model. Quart. J.R. Met. Soc. 106, pp. 747-770.

US National Academy of Sciences, 1975, "Understanding climatic change",
    Washington DC.

Wetherald R.T. and Manabe S., 1975, The effect of Changing the Solar Constant
    on the climate of a General Circulation Model, J. Atmos. Sci., 32,
    pp. 2044-2059.

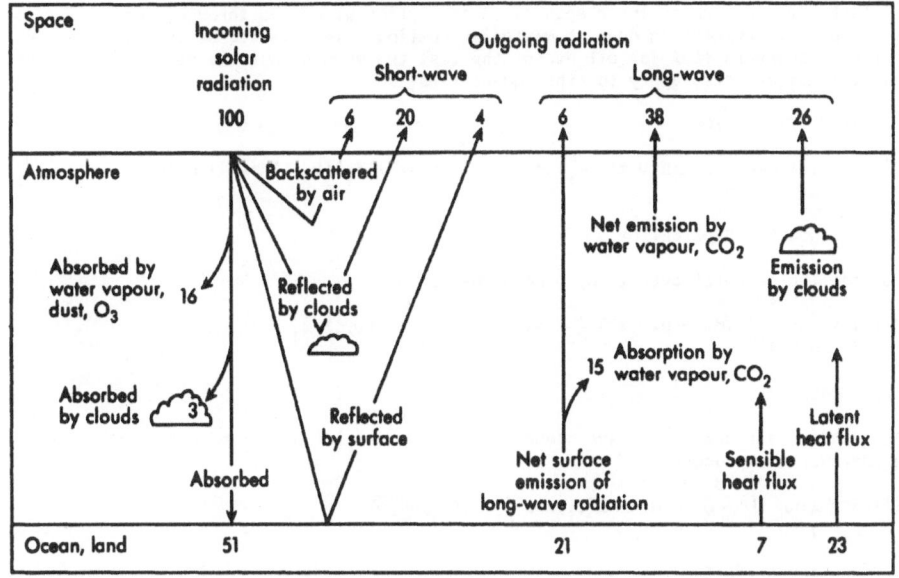

Fig. 1

The radiation balance of the atmosphere-earth system showing the re-
distribution of incoming solar radiation and the mechanism by which these
are affected (from the US National Academy Report, "Understanding Climatic
Change", 1975).

We list here in summary the stages of approximation we can go through from the explicit formulation down to Newtonian Cooling. This is illustrated in terms of downward flux for all except the last two stages, but the same simplifications also apply to other quantities :

EXPLICIT INTEGRATION

$$F^{\downarrow}(z) = \int_0^{\infty} d\nu \int_1^{\infty} \frac{d\mu}{\mu^3} \int_z^{\infty} dz' \; \pi B(\nu,z') \; \frac{d}{dz'} \; \exp\left\{-\mu \int_z^{\infty} \Sigma_i \; S_i(z'') \; f_i \; (\nu,z'') \; p(z'') \; dz''\right\}$$

CURTIS GODSON APPROXIMATION

Replace the integral over z'' by a homogeneous path

$$F^{\downarrow}(z) = \int_0^{\infty} d\nu \int_1^{\infty} \frac{d\mu}{\mu^3} \int_z^{\infty} dz' \; \pi B(\nu,z') \; \frac{d}{dz'} \; \exp\left\{-\mu \sum_i S_i \; f_i \; (\nu,\bar{p}) \; \bar{m}\right\}$$

BAND MODEL

Replace the integral over wave number of the exponential by a band model and a sum over wave number

$$F^{\downarrow}(z) = \Sigma_j \Delta\nu_j \int_1^{\infty} \frac{d\mu}{\mu^3} \int_z^{\infty} dz' \; \pi B(\nu_j,z') \; \frac{d}{dz'} \; T(\nu_j,\mu\bar{m},\bar{p})$$

DIFFUSE APPROXIMATION

Replace the integral over angle ($\mu = \sec\theta$) by a diffusion factor β

$$F^{\downarrow}(z) = \Sigma_j \Delta\nu_j \int_z^{\infty} dz' \; \pi B(\nu_j,z') \; \frac{d}{dz'} \; T(\nu_j,\beta\bar{m},\bar{p})$$

EMISSIVITY

Replace the sum over wave number and the transmission by an emissivity

$$F^{\downarrow}(z) = - \int_z^{\infty} dz' \sigma \Theta^4(z') \; \frac{d\epsilon}{dz'} \; (u(z,z')) \qquad u = \beta \overline{mp} \quad \text{perhaps}$$

COOLING TO SPACE

Ignore all terms in the heating equation, apart from the exchange with space term

$$h(z) = \sigma\Theta^4(z) \; \frac{d\epsilon}{dz'} \; (z, \;)$$

NEWTONIAN COOLING

Linearise cooling to space with respect to temperatures and take $d\epsilon/dz'$ as constant

$$h(z) = a + b\Theta(z)$$

FIGURE 2 :

Stages of approximation in calculating the flux of long wave radiation F (z) at a level z in the atmosphere. The explicit integration at the top may be derived from the theory of radiative transfer. The expressions below are approximations to the explicit integration which become progressively more crude towards the bottom on the figure. (From Rodgers, 1977).

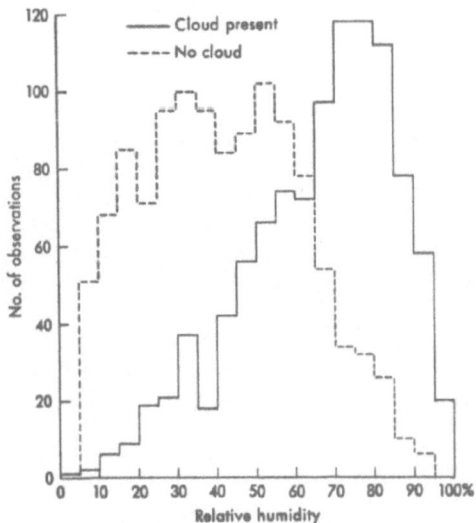

Fig. 3

Frequency distributions of observed mean relative humidity in the layer 700-500 mb. for reports with and without cloud (from Slingo, 1980).

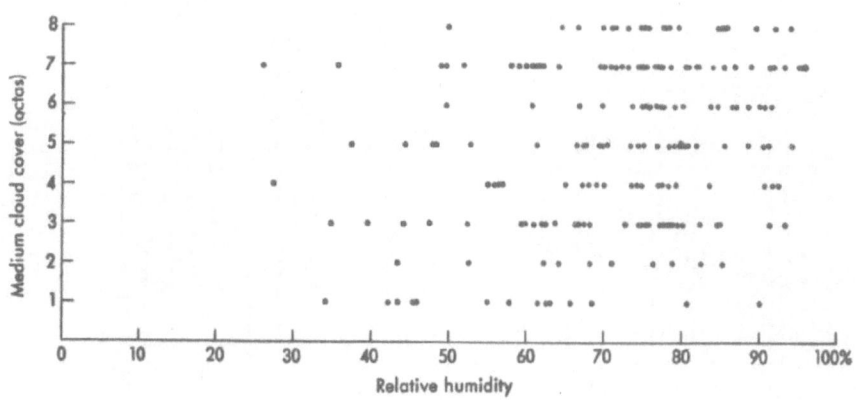

Fig. 4

Scatter diagram of observed medium cloud amount against mean relative humidity for the layer 700-500 mb (from Slingo, 1980).

FIGURE 5(a)

MEAN SEA LEVEL PRESSURE (MB) DECEMBER, JANUARY, FEBRUARY

FIGURE 5(b)

JANUARY PMSL—OBSERVED

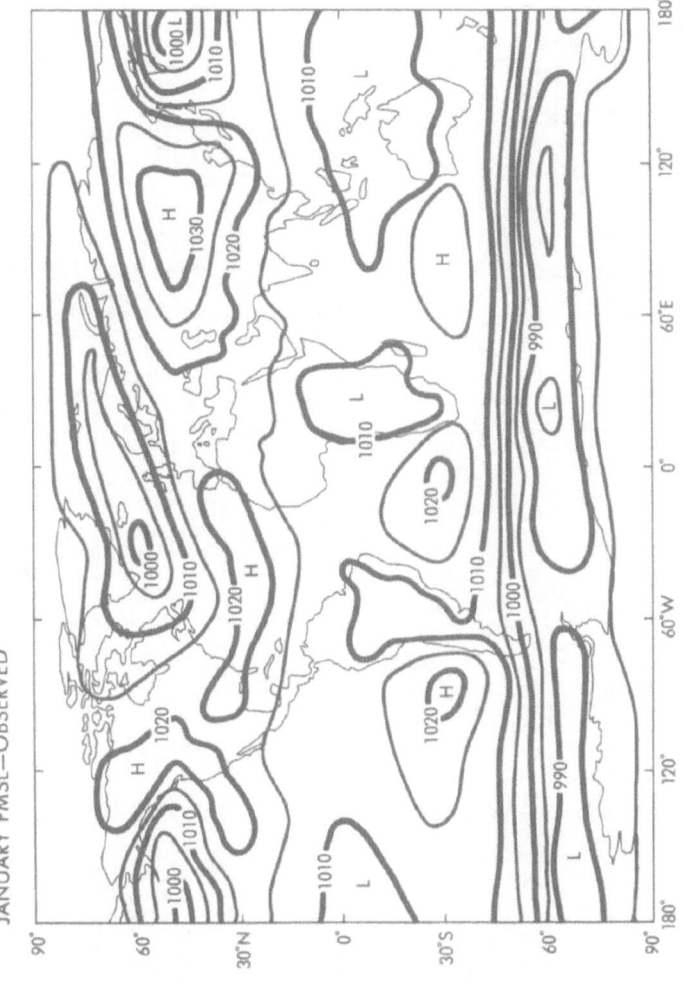

FIGURE 5 Mean sea level pressure. Contours every 5 mbs, heavy contours every 10 mbs.

(a) UK Meteorological Office 5-layer model, averaged over December, January and February.

(b) Observed (January) from Schutz and Gates (1971).

FIGURE 6(a)

Standard deviation of surface pressure (mb)

December, January, February

(Mean of values for each month)

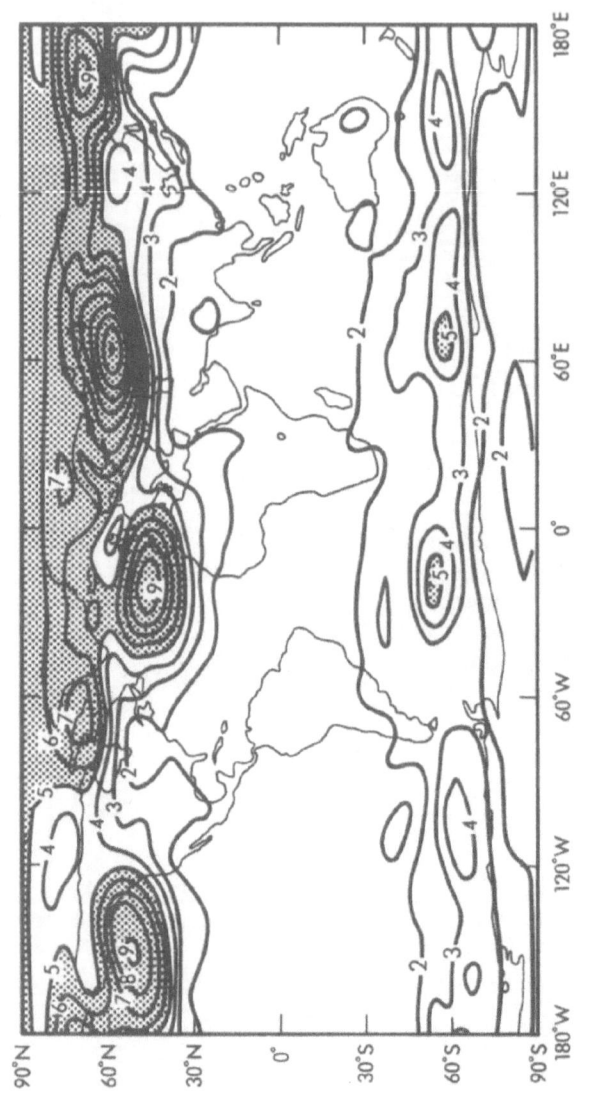

> 5 mb

FIGURE 6(b)

JANUARY STANDARD DEVIATION OF PRESSURE AT MEAN SEA LEVEL 1875—1974 (MILLIBARS)

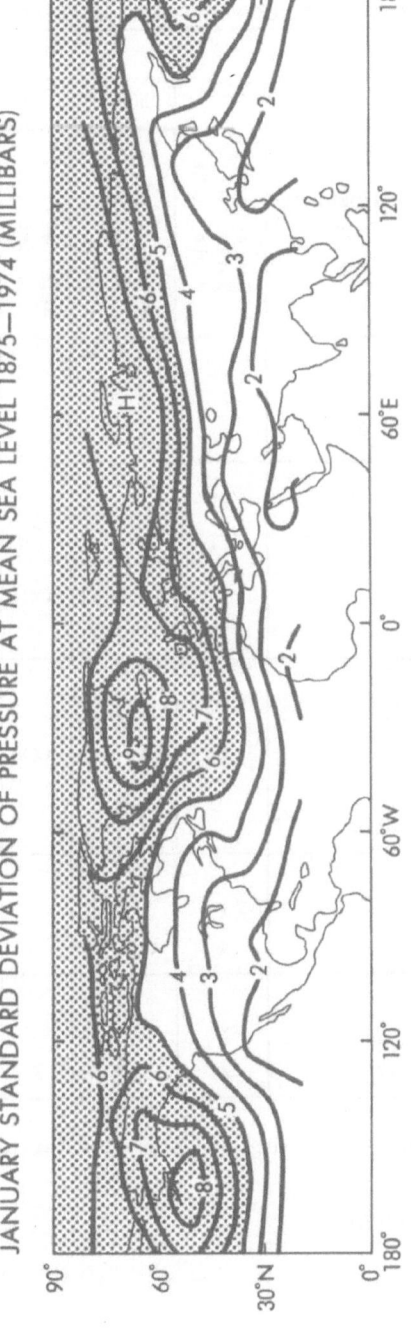

FIGURE 6 Standard deviation of monthly mean surface pressure. Contours every mb,
stippled where greater than 5 mbs.

(a) UK Meteorological Office model, mean of values for December, January
and February.

(b) Observed (January) 1975—1974.

Process and Adjustable Parameters	Parameters Valid Over Whole Globe	Chosen on basis of				
		Tuning expts	Theory	Detailed Models	Observations	Laboratory Experiments
Dymanics						
Diffusion Coeff.	X	X	(X)			
Time Smoothing	X	X				
Radiation						
Gaseous Absorption	X		X	X	X	X
Cloud Amounts	X*	X			X	
Cloud Properties			(X)	(X)	X	
Surface Reflectivity	X		X		X	
Convection						
Parcel Size						
Entrainment and	X	X	X	(X)	X	(X)
Detrainment						
Evaporation of Precipiation						
Boundary Layer						
Drag Coefficient	X		X		X	
Mechanical Mixing	X	(X)	X		(X)	
Other						
Soil Moisture	X	(X)	X		X	
Heat Flux Through Sea Ice	X		X	X	X	X

* Cloud amounts assigned differently in high, low latitudes

x Main sources

(x) Secondary considerations

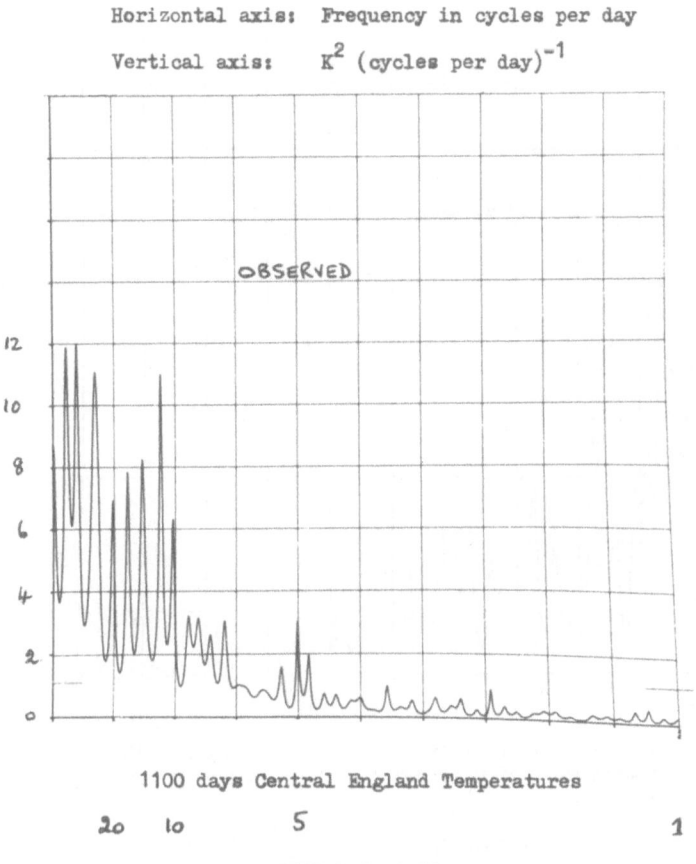

Figure 7. Power spectral density for temperature (with annual and semi-annual cycles removed).

(a) Observed, form Central England Temperatures 1975-77.

Horizontal axis: Frequency in cycles per day

Vertical axis: K^2 (cycles per day)$^{-1}$

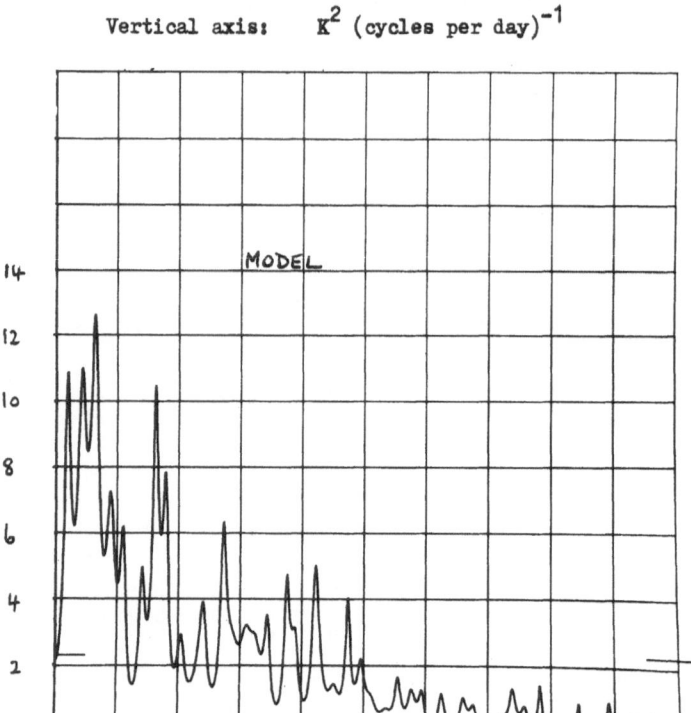

1090 days model **temperatures** near 900 mb **for the Eastern England Grid Box**

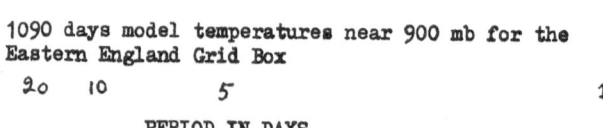

PERIOD IN DAYS

Figure 7. Power spectral density for temperature (with annual and semi-annual cycles removed).

(b) Model temperature at σ = 0.9 (near 900 mbs) for a grid box over Eastern England.

FIGURE 8

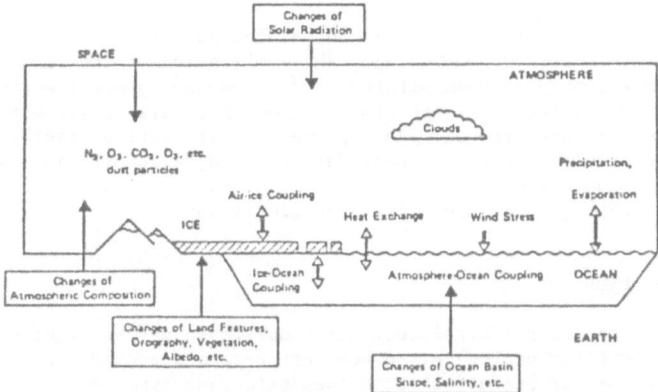

The physical processes and properties that govern the global climate and
its changes (from the US National Academy Report. "Understanding Climatic
Change", 1975).

Application of inverse modelling techniques to palaeoclimatic data

K. Hasselmann
and K. Herterich

Max-Planck-Institut für Meteorologie
Hamburg

Abstract

The method of inverse modelling is summarized and illustrated by examples
from short-term climate modelling. The application of the technique to
palaeoclimatic data is demonstrated by developing a general method for the
construction of linear climate response models simulaneously with the time
calibration of core records. The approach admits full variability of the
time-depth calibration curve under defined integral constraints while
determining the optimal linear climate response to astronomical forcing
consistent with general dynamical side conditions.

1. Introduction

In the last years the palaeoclimatic data base derived from deep-sea
cores and other geological sources has expanded considerably. Continuous
profiles of various climatic indices extending over several hundred
thousand years or longer now exist for a wide distribution of locations.
These data provide not only information on the state of past climates,
but also on the dynamics of the climatic system controlling global climatic
variations in the time scale range from $10^5 - 10^6$ years. However, the
relevant dynamic properties cannot be inferred immediately from the data
records, but must be extracted from a host of other relations involving
the interpretation of climatic indices, the absolute time calibration of
core records, and the form of external climatic forcing, as well as the
various internal interactions within the climatic system. A useful tool
for decomposing such multiple interrelations and establishing the
significance of the diagnostic and dynamical conclusions inferred from the
data is the method of inverse modelling. The technique has become a
standard tool in many areas of geophysics, but has so far found little
application in palaeoclimatology (apart from the climatic interpretation
of proxy data).

In this paper we summarize briefly the basic concepts of inverse modelling,
present a few elementary examples of previous applications in climate
dynamics analysis, and outline some potential applications for palaeoclimatology.

2. Inverse modelling methods

The difference between standard "direct modelling" and "inverse modelling"
is basically minor. Any model designed to simulate an observed set of n
data values $\underline{d} = (d_1, \ldots d_n)$ will normally contain a number m of
adjustable internal parameters $\underline{a} = (a_1, \ldots a_m)$. The modeller will
typically try to "tune" his model through suitable choice of the para-
meters $\underline{a}$ such that the model $\underline{\tilde{d}}$ approximates the observed data $\underline{d}$ as closely
as possible (Fig. 1a). The principal difference between the inverse
modelling approach and standard direct modelling is that the tuning is not

left to the subjective efforts of the modeller, but is automated in the
form of a feedback loop (cf. Fig. 1b).

The automization requires the definition of an error function which is
minimized in the feedback loop. Normally, some quadratic form

$$\epsilon = \sum_{i,j} M_{ij}(d_i-\hat{d_i})(d_j-\hat{d_j}) \tag{1}$$

is chosen where M_{ij} represents a positive definite error metric. The
output of the minimization process then defines a set of optimally
determined model parameters $\underset{\sim}{a}$ as a function of the observed data
$\underset{\sim}{d}$, $\underset{\sim}{a} = \underset{\sim}{a}(\underset{\sim}{d})$. Thus the relation between model parameters and data is in-
verted relative to the direct modelling approach, in which the simulated
data are predicted as output for a given set of input model parameters
$\underset{\sim}{a}$, $\underset{\sim}{d} = \underset{\sim}{d}(\underset{\sim}{a})$.

The simple step of automating the model fitting procedure opens up a
number of possibilities not amenable to subjective tuning procedures :

(1) The models may contain a fairly large number m of free parameters.
The parameters $\underset{\sim}{a}$ may often be related to important physical properties
of the system which could not be easily inferred from the data
without model inversion techniques (cf. next section).

(2) The errors $\delta\underset{\sim}{a}$ induced in the model parameters by errors $\delta\underset{\sim}{d}$ in the
data can be systematically investigated. This enables the assessment
of the statistical sugnificance of the model for a given data error
covariance matrix.

(3) The ability to carry out a quantitative model error analysis and
significance assessment provides further the basis for investigating
hierarchies of models involving an increasing number of free parameters. Typically, the effect of introducing additional parameters
into a model is to increase the "skill" of the model through the
reduction of the error , but at the same time to decrease the
"significance" of the model, as expressed in terms of the model error
covariance matrix $<\delta a_i \delta a_j>$ (cf. Davis, 1977, Barnett and Hasselmann,
1979). The model hierarchy is developed up to a critical order for
which the significance falls below some prescribed acceptance level
(cf. Fig. 2). By systematically increasing the order of the model
until this cut-off point is reached, the maximum content of
statistically significant information can be extracted from the data.

(4) The form (1) of the error function can be readily generalized to
include further side conditions, such as the requirement that the
model should be as smooth as possible, or lie as close as possible to
some favoured theoretical model. Without side conditions, the numbers
of free paramaters characterizing a model class must be restricted to
be smaller than the number of data values. The introduction of
continuous integral side conditions is formally equivalent to the
introduction of a continuum of artificial data. Thus an infinite
dimensional continuum of models may now be considered, without prior
limitation of their functional form. The classical inverse modelling
papers of Backus and Gilbert (1967) and Gilbert (1971) addressed this
general case. The applications to palaeoclimatic data discussed in
section 5 also fall in this class.

A basic limitation of the inverse modelling approach is that the models need to be kept relatively simple, since many cycles through the iteration loop (Fig. 1b) are normally required to minimize the error (particularly for more interesting applications involving a large number of model parameters). Some examples from the field of short term climate modelling (one month to a few years) are given in the next section.

3. Some examples

Inverse modelling techniques are most useful in situations in which a fairly extensive data set for model construction exists, but the physical concepts proposed to explain the data are still relatively rudimentary and can therefore be cast in rather simple models. A typical case is short term climate variability on time scales of months to years. Extensive time series on global climate variability exist for these periods (sea surface temperatures, atmospheric temperatures and pressures, precipitation, etc.), but relatively little is known of the physics governing the observed climatic fluctuations.

An example of the application of inverse modelling for these time scales is given in Fig. 3, which shows the distribution of near surface currents determining the advection of sea surface temperature (SST) anomalies in the North Pacific (Herterich and Hasselmann, 1983). The currents were inferred alone from SST observations by fitting a simple model of the form

$$\frac{\partial T}{\partial t} + u_i \frac{\partial T}{\partial x_i} - \frac{\partial}{\partial x_i} (D_{ij} \frac{\partial T}{\partial x_j}) + \lambda T = n \qquad (2)$$

to 25 years of monthly SST anomaly data for 5° (MARSEN) squares in the area shown. In eqn (2), T denotes the SST anomaly, u_i the two-dimensional current vector, D_{ij} a diffusion tensor, λ a feedback parameter and n a (temporal) white noise forcing term representing the short term fluctuations of the heat transfer across the air-sea interface.

The model was optimized by fitting the simulated output to the measured SST auto- and cross spectra. It yielded fields of all coefficients u_i, D_{ij}, λ and the spectral level and spatial correlation scales of the white noise forcing.

Model fitting to the statistical moments (spectra, covariance functions) of the data rather than the data time series themselves is generally appropriate when the input function driving the model is not given explicitly, but is known only in terms of its statistical properties (e.g. Olbers et al., 1976, Long and Hasselmann, 1979). However, the time series can also be used directly in some cases by adjusting the model coefficients such that the unknown residual input n is minimized, cf. Box and Jenkins (1976). In the present example, this technique was not applied, since the model requirement was not that the input should be white, but not necessarily small. A similar analysis has been applied by Lemke et al. (1980) to investigate the variability of sea ice on monthly to interannual time scales and by Lemke (1977) to determine whether the climate variability in the time scale range $10^2 - 10^6$ years can be explained as the response to stochastic white noise forcing.

A more detailed analysis of the model dynamics is possible if the input function is known explicitly, so that the input and response functions can be correlated. This approach has been used, for example, in the construction of short-term climate linear regression prediction models (cf. Barnett and Hasselmann, 1979, Hasselmann and Barnett, 1981, Davis, 1978). In palaeoclimatic modelling, the method can be applied to determine the dynamical response characteristics of the climate system to astronomical variations of the solar insolation.

In general, however, the construction of palaeoclimatic models will probably need to be based on a combination of the spectral and time series fitting methods, since the driving terms consist of a superposition of the known Milankovitch input and stochastic forcing terms, which can be described only statistically. An additional complication of palaeoclimatic model construction is that the time calibration of palaeoclimatic records is not known and must be determined as part of the model fitting procedure. We consider this problem in section 5 after a brief discussion of current time calibration approaches in the following section.

4. Continuous dating of palaeoclimatic records

Following the pioneering work of Hays et al. (1976), the use of the insolation forcing in the Milankovitch spectral bands of 19, 23 and 41 ky as time reference is generally regarded as the most effective technique for establishing the time calibration of palaeoclimatic records. The time axis is generally tuned subjectively to maximize the spectral coherence between the input and response at the Milankovitch frequencies (cf. Hays et al., 1976, Morley and Hays, 1981). The method makes no use of the information contained in the system transfer functions at these frequencies. Thus the precise choice of climate response signal and input forcing function is immaterial, provided both signals contain significant energy in the frequency bands in question.

Herterich and Sarnthein (1983) have recently applied this approach in a formal inverse modelling framework. However, the inverse method was applied only in a restricted sense, since the calibration function $z = c(t)$ relating the core depth z to time t was allowed to vary only with respect to 5 adjustable parameters (representing 5 dated levels, as compared with about 30 levels which have been varied in subjective tuning methods). In the general technique discussed in the following section, $c(t)$ can be an arbitrary function of time subject only to certain integral constraints.

Various alternative time calibration techniques have been proposed in which assumptions are introduced, independent of the solar insolation input, to interpolate the sedimentation rates between well dated core levels (Shackleton and Opdyke, 1973, Shackleton and Matthews, 1977, Kominz et al., 1979, Sarnthein et al., 1983). For example, von Grafenstein (1982) and Herterich and Sarnthein (1983), related the sedimentation rate to the (highly variable) carbonate concentration. The method was tested by Herterich and Sarnthein (1983) by subsequently correlating the time calibrated climate signal with the solar insolation. Significant coherences were found in the Milankowitch frequency bands, with coherence levels only slightly lower than the values obtained by direct calibration against the solar input. However, the calibration curve CARPOR obtained in this manner deviated significantly, by time separations of the order 20 - 50 kyrs from calibrations obtained from the solar input. Significantly different calibrations were also obtained by the latter technique depending on whether the Brunhes-Matuyama boundary at 730 ky B.P. was regarded as fixed (STUNE) or variable (TUNE) (cf. Fig. 4). We note that all three calibration curves shown in Fig. 4 exhibit statistically significant coherences in the Milankowitch frequency bands at the 90 - 95 % condifende level, although the differences in time calibration are of the order of one to two Milankowitch periods. The calibration ambiguities are related to the property that the total integrated coherence for the three frequency bands passes through a number of relative maxima of comparable magnitude as the calibration curve in systematically varied. Apparently, it is possible to "swallow a period" in regions in which the signals are low without significantly affecting the coherence.

To resolve the ambiguities, additional considerations are needed. One possibility is to penalize calibration solutions which exhibit rapid changes by including a smoothness requirement in the minimized error function. Another approach is to investigate the complex transfer function representing the response of the calibrated climatic signal to the insolation input and test whether the frequency dependencies of the phase and amplitude are consistent with basic dynamical concepts. Fig. 5 shows the transfer functions associated with the three calibration curves of Fig. 4. Based on simple requirements of continuity and causality, the calibration curve TUNE (variable Brunhes-Matuyama boundary) may be rejected as improbable. The calibration curve STUNE (fixed Brunhes-Matuyama boundary) is more consistent with a simple first-order feedback model (cf. section 5) than the carbonate calibrated solution CARPOR. However, the calibrations CARPOR and STUNE should not be regarded as competitive, but rather as complementary : STUNE models the gradual changes of the sedimentation rate, while CARPOR is determined largely by the short time scale fluctuations of the sedimentation rate which are implied by the observed short term fluctuations of the carbonate concentration.

We note that none of the models yield plausible transfer functions at the eccentricity period of 10^5 years, unless a high Q resonance of the climatic system is assumed at this frequency. This is a well=known consequence of the high energy peak in the climatic record at this period which has motivated a number of model constructions exhibiting quasi self-generated oscillations at this frequency.

Ideally, a flexible, general time calibration and dynamical model fitting procedure should attempt to combine the diverse inputs and constraints determining the time calibration and model dynamics into a single optimization algorithme, in which the bestfit time calibration and optimal dynamical model are generated as joint outputs of the optimization procedure. The outline of such an approach is developed in the following section.

5. Application of inverse modelling to palaeoclimatic data

We begin with the time calibration problem. Let us assume there exists a prior calibrated depth-time relation $z = c^0(t)$ based on a finite number of dated levels and estimates of the sedimentation rates (derived, for example, from the carbonate concentration profile). We attempt now to improve on this calibration by tuning the observed climatic response $\eta(z)$ to the known astronomical forcing function $\zeta(t)$.

Since $\xi(t)$ has significant energy only in the three Milankovitch bands centered on the periods 19, 23 and 41 ky, we consider for this purpose only the filtered climatic signal in which all energy is removed except within these bands. To avoid encumbering the notation, we retain the symbol $\eta(t)$ for the filtered climatic signal. For the same reason we consider only a single input and single output function. The extension to the multi-dimensional problem of a number of input functions (cf. discussions in Hays et al., 1976, Berger et al., 1981, Kukla et al., 1981, Bruns, 1981, and Herterich and Sarnthein, 1983) and a number of climatic outputs (representing, for example, different climatic indices at different locations) is of considerable interest for the construction of dynamical models, but is conceptually straight-forward and need not be elaborated here.

In improving the time calibration we wish to stay as close as possible to the original calibration function $z = c^o(t)$ while simultaneously satisfying (again "as well as possible") the following additional constraints : well dated levels $z_i = z_i(t_i)$ should be reproduced; the small scale variability of the deviation $c'(t)$ of the net calibration curve $z = c(t) = c^o(t) + c'(t)$ from the prior calibration $c^o(t)$ should be small; and the transfer functions derived from the time calibrated climatic response should be consistent with general dynamical concepts.

The last condition cannot be formulated without considering the dynamic model fitting problem. We assume here for simplicity that the dynamic response model is linear. Although the basic approach can be extended to nonlinear models, it is no longer possible in this case to restrict the analysis to the Milankovitch forcing frequencies, and the relations become more complex.

For a linear model the (predicted) climatic response $\hat{\eta}(t)$ to the forcing can be represented in the general form

$$\hat{\eta}(t) = \int_{-\infty}^{t} T(t-t')\,\xi(t')dt' \qquad T(\xi) \tag{3}$$

or, in the Fourier domain,

$$\hat{\eta}_\omega = T_\omega \xi_\omega \tag{4}$$

where T, T_ω represent, respectively, the system transfer (Green) function and its Fourier transform

$$T_\omega = \frac{1}{2\pi} \int_{-\infty}^{\infty} T(t)e^{-i\omega t}dt \tag{5}$$

and ξ_ω, $\hat{\eta}_\omega$ are defined as full resolution Fourier transforms over the entire interval $-\tau \le t \le 0$ of the given climatic record,

$$(\xi_\omega,\ \hat{\eta}_\omega) = \frac{1}{\tau} \int_{-\tau}^{0} (\xi,\hat{\eta})e^{-i\omega t}dt \tag{6}$$

(The statical smoothing required below for the auto- and covariance spectra may be regarded formally as obtained by averaging over neighbouring frequencies).

Without any prior information on the dynamics of the climatic system, the only condition which can be placed on T is the causality relation $T(t) = 0$ for $t < 0$ (already implied in the upper bound t of the integral in eqn (3)) or the equivalent Kronig-Kramers relations between the real and imaginary components of the Fourier transform T_ω. However, elementary preconceptions on the structure of the climate dynamics may suggest a simple form for T. For example, the climatic response to stochastic white noise forcing $n(t)$ has often been successfully modelled by a simple first order Markov process of the form

$$\frac{d\eta}{dt} + \lambda \eta = n(t) \qquad (7)$$

with a constant linear feedback factor λ which is represented by a transfer fucntion

$$T(t) = \begin{cases} e^{-\lambda t} & (t > 0) \\ \\ 0 & (t < 0) \end{cases} \qquad (8)$$

or

$$T_\omega = \frac{1}{i\omega + \lambda} \qquad (9)$$

The next step in complexity would be a second-order system, with which a resonance at some prescribed period could be modelled (10^5 years would be a favoured candidate).

In the following, we assume that the most likely structure of the transfer function T^o, of the form (8), (9), say, has been identified on the basis of some prior physical concepts. In general, the transfer function T^o will depend on a number m of adjustable parameters a_i.

The problem of simultaneously optimizing the time calibration and dynamical model structure under the stated constraints may then be formulated as the problem of minimizing the general error function

$$\epsilon = \epsilon_o + g_1 \epsilon_1 + g_2 \epsilon_2 + g_3 \epsilon_3 + g_4 \epsilon_4 \qquad (10)$$

with respect to independent variations of the calibration function c, transfer function T_ω and model parameters a_i, where $g_1, \ldots g_4$ represent suitably chosen weighting factors and the individual error functions are defined as follows :

$$\epsilon_o = (\eta(c(t)) - \hat{\eta}(t))^2 dt = (\eta - T(\xi))^2 dt \qquad (11a)$$

$$= 2\tau \; <|\eta_\omega - T_\omega \xi_\omega|^2 > d\omega \qquad (11b)$$

describes the basic deviation between the model prediction $\hat{\eta}$, after calibration of the time axis, and the observed climatic time series (c(t)) (within the Milankovitch frequency bands; the cornered parentheses denote smoothing over neighbouring frequency bands, within the bands, which is formally redundant in the integral (11b), but is required later in the variational equations);

$$\epsilon_1 = \sum_i (z_i - c_i(t_i))^2 = \sum_i \int \delta(t-t_i)(c(t)-z_i)^2 dt \qquad (12)$$

represents the sum of the time calibration errors at the levels z_i for which the date t_i are well-known;

$$\epsilon_2 = \int (c-c^0)^2 dt = \int (c'(t))^2 dt \qquad (13)$$

denotes the mean square deviation of the net calibration function $c(t)$ from the prior calibration $c^0(t)$;

$$\epsilon_3 = \int (\frac{d^2 c'}{dt^2})^2 dt \qquad (14)$$

expresses the constraint that the deviation c' from the prior calibration function c^0 should be as smooth as possible (ϵ_3 could also be replaced by an integral over the square of the slope rather than the curvature of c'), and, finally,

$$\epsilon_4 = <|\xi_\omega|^2> |T_\omega - T_\omega^0|^2 d\omega \qquad (15)$$

represents the deviation between the optimal empirical model T and the preferred theoretical model T_ω^0.

We note that the basic error expression ϵ_0 depends on both the time calibration and the model, while the errors ϵ_1, ϵ_2, ϵ_3 depend only on the calibration, the error ϵ_4 only on the model. The usual decoupled methods of time calibration and model construction consider only the basic error function. It can be shown (e.g. by inspection of eqn (16), below) that the optimal time calibration problem is not well posed for an arbitrary calibration function $c(t)$ when formulated solely in terms of ϵ_0 (the optimal solution consists of a set of perfect fit segments separated by discontinuities). Thus apart from the inherent attraction of introducing all calibration aspects into a single error function, some form of additional constraint is required to yield a meaningful solution.

The variation of eqn (10) with respect to c' yields as minimum condition for the optimal calibration curve c (or c')

$$\frac{d\eta}{dz} [\eta(c(t)) - \hat{\eta}(t)] + g_1 \sum_i (c(t_i) - z_i)\delta(t-t_i) + g_2 c'(t)$$

$$+ g_3 \frac{d^4 c'(t)}{dt^4} = 0 \qquad (16)$$

with boundary conditions

$$\frac{d^2 c'}{dt^2} = \frac{d^3 c'}{dt^3} = 0 \quad \text{at } t = -\tau \text{ and } 0 \qquad (17)$$

The δ-functions in (16) may be removed by integrating across infinitesimal intervals at t_i, yielding the equation

$$g_3 \frac{d^4 c'}{dt^4} + g_2 c' + \frac{d\eta}{dz}(\eta - \hat{\eta}) = 0 \qquad (18)$$

valid for the intervals between the calibration points t_i with matching conditions

$$g_3 \frac{d^3 c'}{dt^3} \Bigg]_{z_i - \varepsilon}^{z_i + \varepsilon} = -g_1(z_i - c(t_i)) \text{ at the calibration points } t_i \qquad (19)$$

The corresponding minimum condition with respect to variations of T_ω yields

$$T_\omega = \frac{\langle \eta_\omega \xi_\omega^* \rangle + g_4 T_\omega^0 \langle |\xi_\omega|^2 \rangle}{\langle |\xi_\omega|^2 \rangle (1 + g_4)} \qquad (20)$$

(The quadratic mean products may be replaced by the appropriate auto- and cross spectra, from which they differ only by a common normalization factor).

Equation (20) is identical to the usual optimal fit transfer function of linear system theory except for the distorsion towards the favoured model T_ω^0 introduced by the terms proportional to g_4. The appropriate value for the weight g_4 can be derived from statistical maximum likelihood considerations based on the statistical uncertainty of the spectral estimates $\langle |\xi_\omega|^2 \rangle$ and $\langle \eta_\omega \xi_\omega^* \rangle$ and the a priori "likelihood" attributed to the validity of the model T_ω^0 (cf. Savage, 1962). In practice, g_4 will generally be of order $p^{-\frac{1}{2}}$, where p is the number of degrees of freedom of the spectral estimate.

Finally, the variation with respect to the preferred model parameters, a_i yields the relations

$$(T_\omega - T_\omega^0) \frac{\partial T_\omega^{0*}}{\partial a_i} \xi_\omega^2 \rangle d\omega = 0 \qquad (i=1, \ldots m) \qquad (21)$$

The optimal time calibration c, optimal dynamical model T and best fit preferred model T_ω are obtained by simultaneous solution of the equations (17) - (21). The solutions can be constructed iteratively : starting from the reference calibration c^0 as first guesses for c and a suitable choice of model parameter a_i as first guess for T_ω, the associated optimal transfer function T can be determined from (20). This defines a theoretical response $\hat{\eta}$, and the solution of the differential equation (18), with boundary conditions (17) and (19), then yields a first iteration of the calibration function c(t). Similarly, equations (21) determine an iterated set of parameters a_i. The procedure is then repeated, a new T_ω being determined from the new calibration c and new preferred model T_ω^0.

Experience with similar coupled optimal fitting problems in other applications (cf. Herterich and Hasselmann, 1983, Long and Hasselmann, 1979, Olbers et al., 1976) indicates that the convergence of such iterative schemes is normally rather rapid.

As pointed out in section 2, an important aspect of inverse modelling, which we have not been able to consider in this example, is the application of the technique to determine the mapping of data errors or statistical estimation uncertainties into model errors. We have also not discussed possible extensions of the model to include stochastic white noise forcing, an important aspect in trying to model the complete palaeoclimatic record (cf. Kominz et al., 1979).

However, even without these extensions the inverse modelling technique described here determine a general optimal time calibration function and dynamical model, subject only to integral constraints which can be clearly specified, and whose effects can be systematically explored by numerical experimentation. The application of these techniques as a diagnostic tool to investigate climate variability in the time scale range $10^3 - 10^6$ years will become indispensible as more palaeoclimatic time series become available for analysis.

References

Backus, G.E., and J.F. Gilbert, 1967. Numerical applications of a formalism
  for geophysical inverse problems, Geophys. J.R. Astron. Soc. 13,
  247-276.

Barnett, T.P., and K. Hasselmann, 1979. Techniques of linear prediction, with
  application to oceanic and atmospheric fields in the tropical pacific,
  Reviews of Geophys. and Space Physics 17, 949-968.

Berger, A., J. Guiot, G. Kukla, and P. Pestiaux, 1981. Long-term variations
  of monthly insolation as related to climatic changes, Geologische
  Rundschau 70, 748-758.

Box, G.E.P., and G.M. Jenkins, 1976. Time Series Analysis, Forecasting
  and Control, Holden-Day, San Francisco, Calif.

Bruns, T., 1981. Unpubl. Diplomarbeit (Master Thesis), Universitaet Hamburg,
  71 pp.

Davis, R.E., 1977. Techniques for statistical analysis and prediction of
  geophysical fluid systems, Geophys. Astrophys. Fluid Dyn. 8, 245-277.

Davis, R.E., 1978. Predictability of sea level pressure anomalies over the
  North Pacific Ocean, J. Phys. Oceanogr. 8, 233-246.

Gilbert, J.F., 1971. Ranking and winnowing gross earth data for inversion
  and resolution, Geophys. J.R. Astron. Soc. 23, 125-128.

von Grafenstein, R., 1982. Unpubl. Diplomarbeit (Master Thesis),
  Universitaet Kiel, 67 pp.

Hasselmann, K., 1979. Linear statistical models. Dyn. Atmos. Oceans 3,
  501-521.

Hasselmann, K., and T.P. Barnett, 1981. Techniques of linear prediction for
  systems with periodic statistics, J. Atm. Sci. 38, 2275-2283.

Hays, J.D., J. Imbrie, and N.J. Shackleton, 1976. Variations in the Earth's
  orbit : pacemaker of the ice ages. Science 194, 1121-1132.

Herterich, K., and M. Sarnthein, 1983. Brunhes time scale : tuning by rates
  of calcium-carbonate dissolution and cross spectral analyses with
  solar insolation, Proceedings of Conference "Milankovitch and Climate",
  Nov. 30 - Dec. 4, 1982, Palisades, New York, U.S.A.

Herterich, K., and K. Hasselmann, 1983. Extraction of sea surface temperature
  advection, relaxation and atmospheric forcing parameters from the
  statistical analysis of North Pacific SST anomaly fields (in preparation).

Kominz, M.A., G.R. Heath, T.L. Ku, and N.G. Pisias, 1979. Brunhes time
  scales and the interpretation of climatic change. Earth Plan. Sci.
  Lett. 45, 394-410.

Kukla, G., A. Berger, R. Lotti, and J. Brown, 1981. Orbital signature of
  interglacials, Nature 290, 295-300.

Lemke, P., 1977. Stochastic climate models, Part 3. Application to zonally-averaged energy models. Tellus 29, 385-392.

Lemke, P., E.W. Trinkl, and K. Hasselmann, 1980. Stochastic dynamic analysis of polar sea ice variability, J. Phys. Oceanogr. 10, 2100-2120.

Long, R.B., and K. Hasselmann, 1979. A variational technique for extracting directional spectra from multi-component wave data, J. Phys. Oceanogr. 9, 373-381.

Morley, J.J., and J.D. Hays, 1981. Towards a high-resolution, global deep-sea chronology for the last 750,000 years. Earth Plan. Sci. Lett. 53, 279-295.

Olbers, D.J., P. Müller, and J. Willebrand, 1976. Inverse technique analysis of a large data set, Phys. Earth Planet. Inter. 12, 248-252.

Savage, L.J., 1962. The Foundations of Statistical Inference, Methuen, London.

Shackleton, N.J., and N.D. Opdyke, 1973. Oxygen isotope and palaeomagnetic stratigraphy of Equatorial Pacific cores V28-238 : Oxygen isotope temperatures and ice volumes onf a 10^5 year 10^6 year scale, Quart. Res. 3, 39-55.

Shackleton, N.J., and R.K. Matthews, 1977. Oxygen isotope stratigraphy of Late Pleistocene coral terraces in Barbados. Nature 268, 618-620.

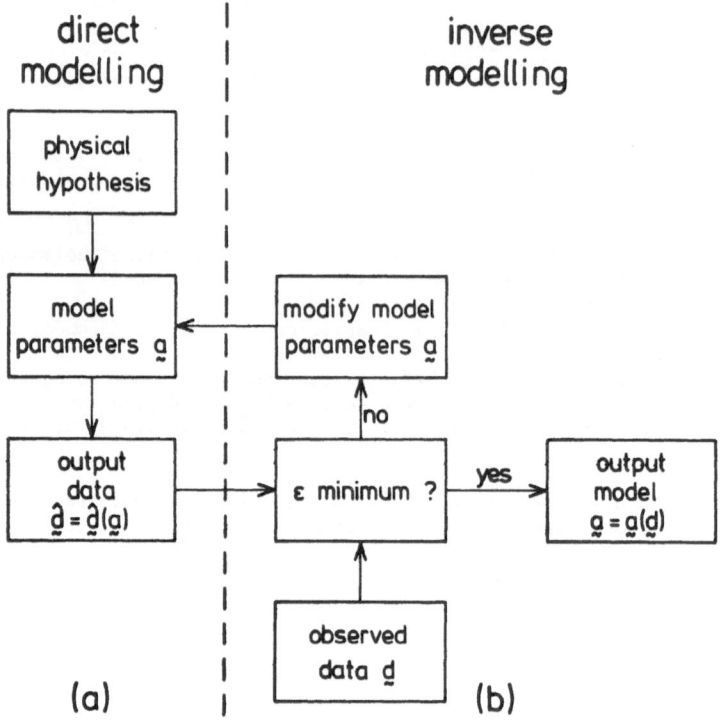

Fig. 1 Relation between direct and inverse modelling methods.
   In the inverse modelling approach, the optimally tuned model is found
   by a closed iteration loop (for very simple, e.g. linear models, the
   inversion can sometimes be given explicitly).

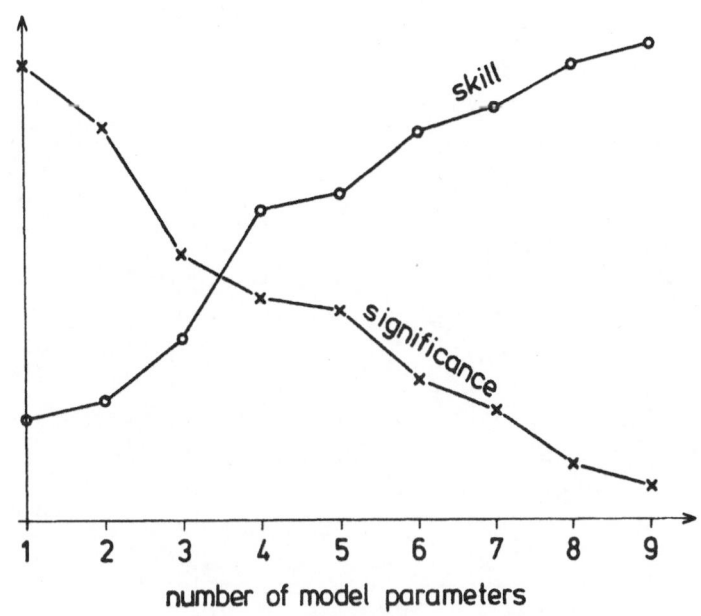

Fig. 2 Contrary dependencies of model skill and significance on number of adjustable model parameters.

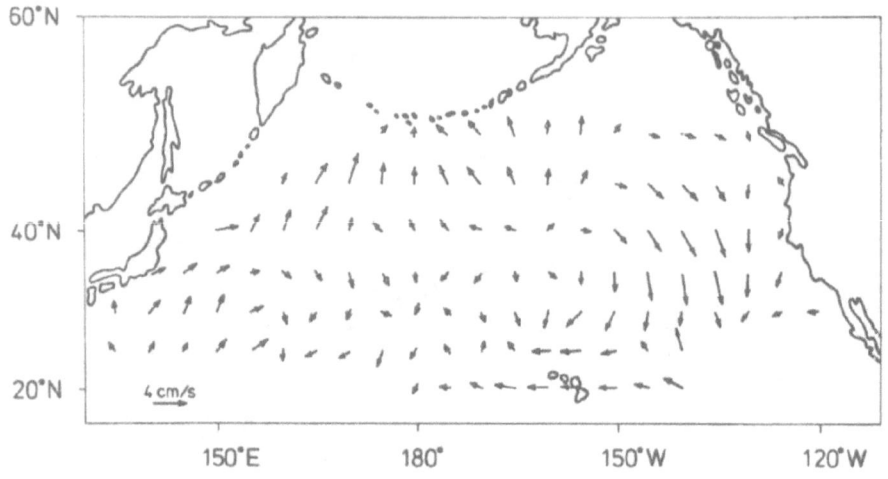

Fig. 3 Field of upper layer ocean currents determining advection of SST as
       inferred by statistical analysis of SST anomaly time series (from
       Herterich and Hasselmann, 1983).

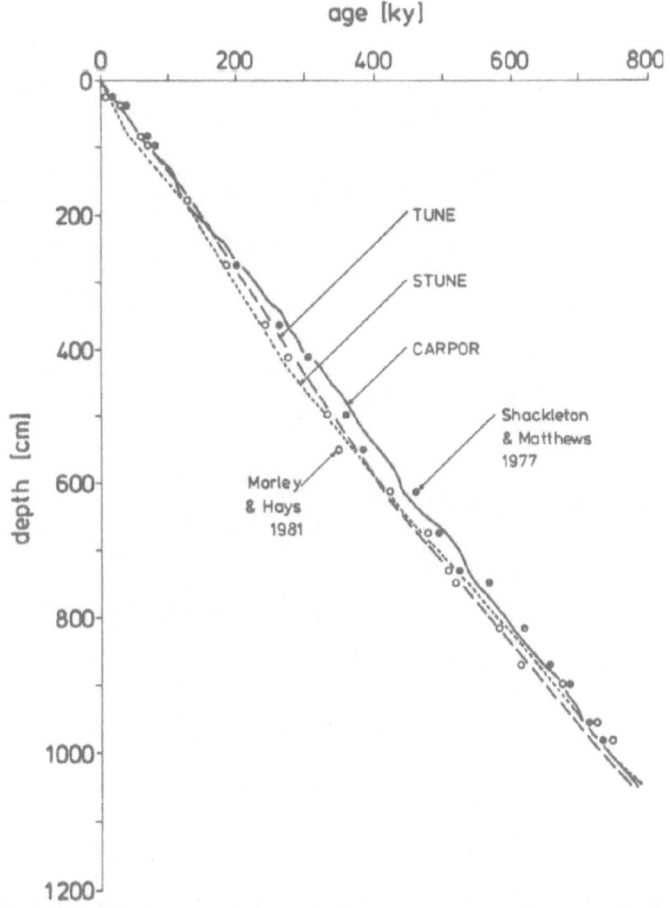

Fig. 4 Time calibration curves for METEOR core 13519 inferred from carbonate concentrations (CARPOR) and cross correlation with July insolation at 65°N with fixed (STUNE) and variable (TUNE) Brunhes-Matuyama boundary (from Herterich and Sarnthein, 1983).

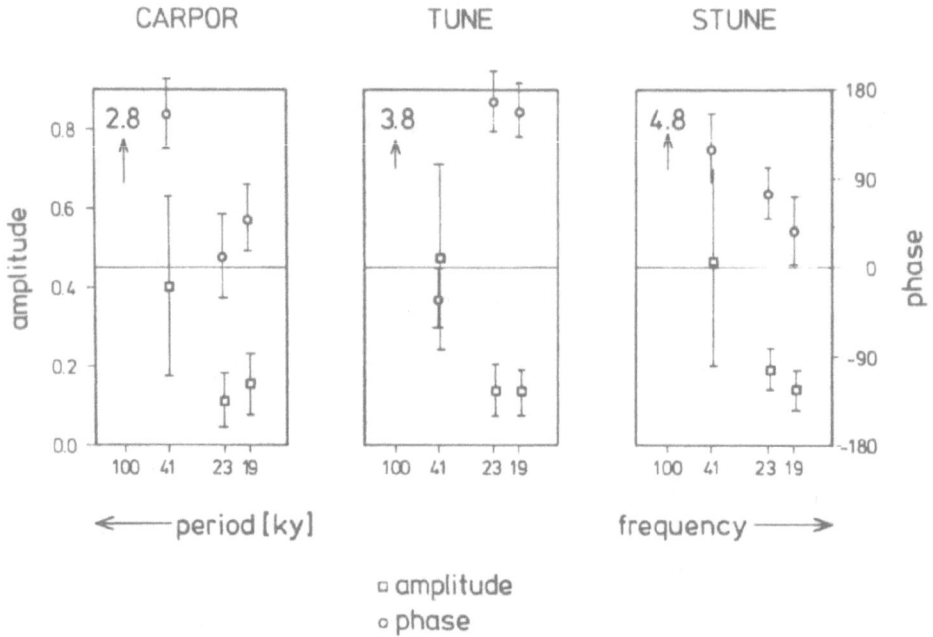

Fig. 5 Transfer functions (amplitude and phase) relating climate response
(δ¹⁸O) for METEOR core 13519 to July insolation at 65°N for the three
time calibrations CARPOR, STUNE and TUNE.

Accuracy of palaeoinsolation and stability in the frequency domain

A. BERGER

Institut d'Astronomie et de Géophysique
Université Catholique de Louvain
2 Chemin du Cyclotron
B-1348 Louvain-la-Neuve
Belgique

Among all the indices against which the geological data are tested to validate the astronomical theory of palaeoclimates, the most popular ones are certainly the orbital elements of the earth and linear or non-linear combinations of them.

However, as the climate system is thermally driven by solar insolation, there is a real interest to check whether or not there are some relationships between insolation parameters and the climate at the global scale. The difficulty which then first arises is to determine which are the most realistic latitudes and during which periodes of the year. The adjustment may be tested in the time and/or the frequency domains. This is why the different kinds of insolation which are supposed to be used for modelling the climate or for simulating the climatic variations, are carefully reviewed (computation, accuracy and spectrum).

These insolation parameters are : half year astronomical seasons (length, total and mean insolation), half year caloric seasons, astronomical and meteorological seasons (length, total and mean insolation), monthly mean insolation, mid-month (solar date) and calendar date insolations.

These insolations depend upon (i) 4 astronomical parameters which all affect the total energy received by the planet : the astrophysical solar constant, the lenght of the ropical year and of the day, the secularly variable mean distance from the Earth to the Sun (defined by the eccentricity, e), (ii) and 2 other ones which re-distribute differently the energy among the latitudes and months : the climatic precession, e sin , and the obliquity, . This dependance is summarized in table 1.

Table 1 - Insolation as a function of astronomical parameters
(++ means stronger dependancy)

Mid-month insolation at equinox		+
at soltstice	+	++
Half-year astronomical seasons		
- total insolation	+	
- lenght		+
- mean in polar latitudes	++	+
in equatorial latitudes		+
Caloric seasons polar latitudes	+	
equatorial latitudes		+
Meteorological seasons (astronomical definition)		
- total insolation	+	
- lenght		+
Meteorological seasons (monthly mean)	+	++

The accuracy of the long-term variations of the astronomical elements and
of the insolation values and the stability of their spectrum have also been
analysed by comparing 7 different astronomical solutions and 4 different
time spans (0-0.8 Myr BP, 0.8-1.6 Myr BP, 1.6-2.4 Myr BP and 2.4-3.0 Myr BP).
The general conclusions are, for the accuracy in time, that improvements are
necessary for periods further back than 1.5 Myr BP. About the stability of
the frequencies, the fundamental periods (around 40, 23 and 10 Kyr) do not
deteriorate with time over the last 5 Myr but their relative importance for
each insolation and even astronomical parameter is a function of the period
considered.

Session A : Abrupt Climate Changes

ICE CORE INDICATIONS OF ABRUPT CLIMATIC CHANGES

W. Dansgaard,
Geophysical Isotope Laboratory, University of Copenhagen,
Haraldsgade 6, Copenhagen 2200, Denmark.

H. Oeschger,
Physikalisches Institut, University of Bern, Switzerland.

C.C. Langway,
Geological Department, University of New York at Buffalo, N.Y.

Polar ice sheets are proving richer sources of information about past atmospheric conditions than any other sequency of Quaternary sediments. All kinds of fall-out from the atmosphere, such as air-borne continental dust and biological material, volcanic debris, sea salts, cosmic particles, and cosmic ray produced isotopes, are deposited on the ice sheet surfaces along with the snow. The snow pack is gradually compressed into solid ice with small cavities containing samples of atmospheric air. In the coldest areas of the ice sheets, the impurities remain in the ice as indicators of the chemical and physical conditions of the atmosphere at the time of deposition.

A new deep ice core has been recovered from Dye 3 in South Greenland. Several chemical and radio-isotope profiles are being established along the core. The CO_2 and ^{10}Be analyses are particularly promising.

Continuous profiles of $\delta(^{18}O)$, acidity and concentration of continental dust are of fundamental importance for absolute dating and climatic interpretation.

Comparison of the $\delta(^{18}O)$ profile with that from Camp Century and with a deep sea foraminifera record suggests that (i) the new core reaches back to ca. 90,000 years B.P. in a continuous sequence; (ii) the general trends reveal all of the relevant Emiliani stages recorded in deep sea cores; (iii) the details in the Weichelian part of the ice core records are probably climatically significant.

Numerous abrupt changes in δ and dust concentration occurred simultaneously in SE and NW Greenland from 75,000 to 10,000 yrs. B.P. They may be ascribed to shifts between two quasi-stationary modes of atmospheric circulation in the North Atlantic. Surges of the entire ice sheet would lower the surface elevation, and the consequent rise in sea level might cause flooding of vast dry areas on the continental shelves. Hence, surges would seem to be an alternative explanation for the shifts from low to high δ's, which are accompanied by abrupt decreases in the dust concentration. But, the quite different ice flow conditions in SE and NW Greenland make it difficult to envisage how the entire ice sheet could surge at the same time, unless these events were provoked by much more drastic surges of the Antarctic ice sheet ? The δ profile along the Vostok ice core does not exclude this possibility, but the profile is not yet detailed enough to verify it. However, if the surge hypothesis is adopted, the Greenland shifts from high to low δ values must be explained by growth of the ice sheet, and this process is too slow to be plausible.

The Camp Century $\delta(^{18}O)$ record suggests a dramatic termination of the Eemian interglacial, possibly followed by a period of intermediate temperatures and high accumulation rates.

The onsets of several glaciations coulb be studied in details on deep ice cores drilled in a favorable areas. Techniques have been developed to establish time series of many parameters, e.g. accumulation rate, surface temperature, volcanic activity, CO_2 concentration in the atmosphere, humidity of marine air masses, and cosmic radiation flux.

Possible future deep ice core drilling should be done in areas of :

1. simple ice flow history, to simplify ice flow modelling,

2. high accumulation rate, to ensure continuity of the time series, and

3. no melting at bedrock, to ensure a long time range, nor at surface, to ensure undisturbed stratigraphy and composition of the air included in the bubbles in the ice.

Abrupt Climate Changes : The terrestrial record
W. WATTS
University of Dublin, Ireland

LATEGLACIAL VEGETATION AND CLIMATE OF THE ATLANTIC COAST OF EUROPE

(a) IRELAND

The first investigation of Lateglacial sediments in Ireland and Britain was carried out by Jessen and Farrington (1938) at Ballybetagh near Dublin. They showed that organic lake-mud which contained remains of giant deer (Megaloceros giganteus) was sealed by stony inorganic sediments, believed to be formed by solifluction. The interpretation was that a period of relatively temperate climate was brought to an end by a phase of arctic cold. Comparison was made with the Lateglacial stratigraphy of Denmark at the classic Allerød site. Subsequently, Jessen (1949) developed a general scheme of Lateglacial events in Ireland (Table I) and correlated them with events in Denmark.

Jessen's simple tripartite scheme with a 'warm' Allerød interstadial, sandwiched between two cold phases with inorganic sediments, was generally accepted until recently when more detailed pollen analysis revealed that the Lateglacial was more complex than had been realised. In particular, his failure to identify juniper, an abundant and important component of the vegetation during part of the Lateglacial, was a serious weakness. Jessen's relatively primitive pollen analysis was balanced by the meticulous attention he and his pupil Mitchell (1953, 1954) gave to the determination of plant macrofossils which proved to be a rich and invaluable source of information for ecological and plant-geographical interpretation.

With the application of more detailed pollen analysis, it became clear that Jessen's scheme was no longer adequate. Changes in the vegetation cover did not necessarily correlate exactly with changes in sediment type (Watts, 1963) and there were too many distinct events in the vegetation and climatic history of Ireland to be fitted into three units (Singh, 1970; Smith, 1970).

It is certain that the climate and vegetation of the phases of the Lateglacial were complex, with much variation between sites, depending on soil type and elevation (Watts, 1977, 1980). Table II briefly resumes the vegetational events and chronology.

The lithostratigraphy is also variable from site to site. At many localities the total thickness of sediment is less than 100 cm, but some very thick sequences are known. The deepest part of the Coolteen basin (Craig, 1978) has up to 800 cm of Lateglacial sediment, other sites have between 400 and 500 cm. Such rapid accumulation is scarcely encountered elsewhere in northwest Europe (Pennington, 1977). It means that Ireland is particularly favoured for detailed stratigraphic and climatic study of the Lateglacial. The oldest Lateglacial sediments are usually fine silts and clays with no countable pollen, or pollen at a very low concentration. These are rock-flour deposits of decaying ice-sheets or erosional sediments washed into basins from unvegetated mineral soil surfaces. Sometimes thicknesses of several metres of such material are encountered, consistent with the estimate of Ruddiman and McIntyre (1981) that rapid melting was taking place from as early as 16,000 yr bp until a warming tendency at 13,000 yr bp.

Clays and silts are followed by progressively more organic deposits, often calcereous. At Coolteen black organic muds with up to 50% loss on ignition values early in the Lateglacial correspond with the first phase of

Table I. Jessen's scheme for the Lateglacial in Ireland

		Denmark
Ireland		
Younger Salix herbacea Period (Zone III)	– tundra-like vegetation with sub-arctic heaths, solifluction	Younger Dryas Period
Lateglacial Birch Period (Zone II)	– birch copses, open herb-rich vegetation	Allerød Period
Older Salix herbacea Period (Zone I)	– tundra-like vegetation, solifluction	Older Dryas Period

juniper abundance. This may have been the time with the warmest climate in the Lateglacial. The organic or carbonate content of sediments was high, and diverse aquatic plants such as Ceratophyllum demersum (hornwort) and several Potamogeton (pondweed) species flourished. The algal flora was unusually diverse (Watts, 1977). These and other lines of evidence, reviewed by Watts (1977) and Craig (1978), suggest that the First Juniper Phase had an exceptionally favourable climate, not equalled until the Postglacial.

Coope (1977) has presented evidence from fossil beetles to show that a warm interstadial (The Windermere Interstadial) took place in northwest Britain in the closing stages of the Last (Devensian) Glaciation before 13,000 bp, but later than 14,000 bp. The climate is considered to have been as warm in summer as it is today and with winter temperatures a little lower than at present. A cooling took place about 12,200 bp and a period of over one thousand years (the Allerød oscillation) followed, during which summers were about 3°C cooler than during the thermal maximum. The dates proposed by Coope are somewhat older than those suggested here for Ireland, but the similarity of the analysis of the climatic sequence from both plant and beetle evidence is very striking and at variance with the accepted view from continental Europe (Iversen, 1954) which tends to equate the Bølling and Allerød climates and does not distinguish an early warm phase so clearly.

The earlier part of the Lateglacial vegetation sequence is like an interglacial cycle (Iversen, 1958) and one would expect invasions of more demanding pioneering trees and shrubs, such as tree birches, tree willows and aspen to follow. However, a brief period of erosion between about 12,000 and 11,800 yr bp appears to have broken the cycle. At many sites, especially on the west coast of Ireland, a conspicuous layer of silt is found after an earlier period of juniper expansion. In the Dublin area, Hippophae (sea-buckthorn), a plant of unstable soils, appeared briefly and juniper itself fell to low values once more. At Coolteen high productivity and high juniper pollen concentrations ended simultaneously. The climatic deterioration caused soil movement and decline in juniper. The latter may have been caused by a lowering of both temperature and precipitation, for juniper is killed by wind in the sub-arctic unless protected by snow cover (Iversen, 1954). The end of the First Juniper Phase therefore shows a serious climatic deterioration and erosion of the upland, marked by silt movement into lake basins. It is maintained that this event is similar in climatic character to the Younger Dryas event, though smaller in scale, and probably had the same causes.

The Grass Phase occupied about 1,000 years. In eastern Ireland the sediments became more inorganic and the early diversity of algal species was lost. The received opinion is that the Grass Phase (in part, Jessen's Zone II or Allerød) was the 'warmest' part of the Lateglacial. This seems unlikely to be true. It is more probable that it was a relatively cold time with moderate erosion or deflation from herb-dominated vegetation that cannot have formed a complete cover. It may have been steppe-like in character with trees or shrubs confined to protected places. Thus the environment for giant deer an reindeer may have been tundra-like grassland, well-suited to cold-adapted grazing and browsing mammals.

At the end of the Grass Phase severe erosion began once more, marking the beginning of the Younger Dryas Period. The sediments are predominantly inorganic and may contain stones. Leaves of dwarf willow are often abundant. This is consistent with solifluction and late snow-beds. The distinctive pollen flora and often sparse pollen grains suggest an incomplete vegetation cover of arctic species.

Table II. Subdivisions of the Lateglacial

GENERAL SCHEME FOR NORTHWEST EUROPE			IRELAND		
Mangerud et al., (1974)					
Flandrian	10,000 yr bp - present		Postglacial	10,000 yr bp - present	
	Younger Dryas	11,000-10,000		Artemisia Phase (Younger Dryas)	10,900-10,000
	Allerød	11,800-11,000		Grass Phase	11,800-10,900
Late Weichselian	Older Dryas	12,000-11,800	Lateglacial	Erosion ca.	12,000-11,800
	Bølling	13,000-12,000		First Juniper Phase	12,400-12,000
	?	? -13,000		Rumex Phase	13,000-12,400

Mitchell (1973) has suggested that pingos, which are abundant in southern Ireland outside the end-moraine of the Last Glaciation, were still active during the Younger Dryas. Investigations up to the present support that view, but there are still too few site studies to confirm it strongly. At Lough Nahanagan, a mountain lake south of Dublin (Colhoun and Synge, 1980; Watts, 1977), there is clear evidence that a glacier formed in the Younger Dryas at an elevation of 450 m and moved from the back wall of the cirque to form a substantial moraine. This incorporates blocks of organic sediment dated to 11,600 yr bp with an early Lateglacial pollen assemblage. The depression of the snowline to permit a new glacial advance at Lough Nahanagan represents a considerable lowering, calculated at 7.2°C, of the average annual temperature in comparison with today.

Open-system pingos, apparently the type found fossil in Ireland and Wales (Mitchell, 1973; Watson, 1971), occur at present in the discontinuous permafrost zone of Central Alaska, with mean annual temperatures as high as -1°C to -2°C (Washburn, 1973). If the pingos were active in the Younger Dryas, at least discontinuous permafrost was present. The advance of cirque glaciers and the evidence for widespread solifluction point to a very cold climate, which has not been quantified satisfactorily as yet in Ireland. None of the phenomena observed require high precipitation. Ruddiman and McIntyre (1981) observe that the eastern North Atlantic, though very cold, was not frozen over in the Younger Dryas. This would favour transport of moisture from the oceans to form precipitation, so that it is unlikely that the cold phase approached aridity.

The cause of the Younger Dryas climatic deterioration have been clarified by recent studies of ocean cores (Ruddiman and McIntyre,1981). After 13,000 yr bp the North Atlantic supported a diverse subpolar fauna. The amount of meltwater entering the ocean and iceberg activity are assumed to have been reduced. Between 11,000 and 10,000 yr bp an exclusively polar fauna appeared in the cores. A fall in ocean surface temperature of 10°C is assumed. The Younger Dryas Period on land correlates with southward advance of the Polar Front. The reappearance of cold oceanic water off the west coast of Ireland was directly responsible for the Younger Dryas climatic deterioration. The earlier deterioration in the Erosion Phase and the maintenance of a cold climate during the Grass Phase suggest that the Polar Front had reversed its retreat already by 12,000 bp and may have readvanced slowly over the next thousand years with a further rapid and severe deterioration at the beginning of the Younger Dryas Period. In general, authors have tended to blur the distinctions between the sub-divisions of the Lateglacial. It is now clear that an early warm period (the First Juniper Phase) was succeeded by a colder Grass Phase and then by a very much colder Younger Dryas Period. Studied with careful attention to stratigraphy, it becomes clear that the major units have quite distinctive floras, each with distinct ecological and climatic implications that still require quantification.

(b) SPAIN AND PORTUGAL

The early studies of the Lateglacial in Spain are due to Menéndez Amor and Florschütz (1961). At a marsh near Sanabria in northwest Spain it was shown that sedimentation began before 13,700 bp. The details of this work cannot be fully confirmed in recent studies at the same site carried out as part of the EEC Climatology Programme. In a new core of 10 m length the base is dominated by pollen of juniper, Artemisia and Ericales. This is followed by a peak of birch pollen and then by pine with oak. A secondary peak of birch is associated with a recession of pine and oak. This gives the appearance of a successional process after deglaciation. Possibly the second birch peak represents the Younger Dryas Period. Radiocarbon dates are awaited. At present it can be stated that there is considerable divergence between the older and the more recent work. The applicability of the climatic/ vegetational sequence identified in Ireland is at present uncertain. Plant macrofossils show early invasion of tree birch and of pine. The vegetation always contained heath plants, abundantly represented among macrofossils from the base of the core. A newly investigated site, Lago de Ajo, also shows a successional process with a long phase with herbaceous pollen. This site lies at about 2,000 m in the Cantabrian Mountains of northwest Spain. An early peak of birch pollen associated with macrofossils of tree birch is dated to before 14,000 bp. Radiocarbon dates wich are available from this site also do not allow an exact comparison with the Lateglacial sequence known from Ireland. The evidence is for early deglaciation before 14,000 bp.

Studies are also in progress at Lagoa Comprida in central Portugal where a herb-dominated flora with Artemisia also opens the sequence. This site is at high elevation in the granitic mountains of the Serra da Estrela of central Portugal. All three sites are cirque or iceblock hollow lakes where ice of the Last Glaciation has melted out. The presence of large local glaciers in northwest Spain and central Portugal is noteworthy.

(c) DISCUSSION

The studies carried out under the EEC Climatology Programme in Ireland and the Iberian Peninsula have not yet progressed to a point where a full evaluation can be made. The following considerations appear to be important :

1. The vegetation sequences in Portugal, Spain and western Ireland are not sufficiently similar to one another that a common pattern of events is obvious. Correlation of the pollen diagrams must depend on a substantial number of radiocarbon dates rather than interpretations of the significance of particular pollen curves. Attempts will be made to obtain a high resolution chronology for the events in both areas.

2. The translation of pollen data into climatic data depends on the completion of a programme of surface samples and short sediment cores obtained in both regions. It will be difficult to obtain any modern analogue for the Artemisia-dominated vegetation of the early Lateglacial in the Iberian Peninsula. It appears to indicate a rather arid climate which was also cold enough to allow the establishment of a small ice-cap or a complex of substantial glaciers as far south as central Portugal.

REFERENCES

Colhoun, E.A. and Synge, F.M. (1980). Proc. R. Ir. Acad. B80, 25-45.

Coope, G.R. (1977). Phil. Trans. R. Soc. B280, 313-340.

Craig, A.J. (1978). J. Ecol. 66, 297-324.

Iversen, J. (1954). Danm. geol. Unders. Raekke II, Nr. 80, 87-119.

Iversen, J. (1958). Uppsala Univ. Arsskr. 6, 210-215.

Jessen, K. and Farrington, A. (1938). Proc. R. Ir. Acad. B44, 205-260.

Menéndez Amor, J. and Florschütz, F. (1961). Estudios Geologicos 17, 83-99.

Mitchell, G.F. (1953). Proc. R. Ir. Acad. B52, 225-281.

Mitchell, G.F. (1954). Danm. geol. Unders. Raekke II, Nr. 80, 73-86.

Mitchell, G.F. (1973). Proc. R. Ir. Acad. B73, 269-282.

Ruddiman, W.F. and McIntyre, A. (1981). Palaeogeog., Palaeoclimatol., Palaeoecol. 35, 145-214.

Singh, G. and Smith, A.G. (1973). Proc. R. Ir. Acad. B73, 1-51.

Smith, A.G. (1970). In "Studies in the Vegetational History of the Britisch Isles" (D. Walker and R.G. West, eds), pp. 81-96. Cambridge University Press.

Washburn, A.L. (1973). "Periglacial Processes and Environments". Edward Arnold, London.

Watson, E. (1971). Geol. J. 7, 381-392.

Watts, W.A. (1963). Ir. Geogr. 4, 367-376.

Watts, W.A. (1977). Phil. Trans. R. Soc. B280, 273-293.

Watts, W.A. (1980). In "Studies in the Lateglacial of North-West Europe" (J.J. Lowe, J.M. Gray and J.E. Robinson, eds), pp. 1-22. Pergamon Press, Oxford.

EVOLUTION CLIMATIQUE DE LA MEDITERRANEE ORIENTALE AU COURS DE LA DERNIERE DEGLACIATION

par :

Wladimir D. NESTEROFF [1]
Colette VERGNAUD GRAZZINI [1]
Laure BLANC-VERNET [2]
Philippe OLIVE [1]
Jacqueline RIVAULT-ZNAIDI [1]
Martine ROSSIGNOL-STRICK [3]

(1) Département de Géologie Dynamique. Université Pierre et Marie Curie, 4, Place Jussieu, 75230 PARIS Cédex 05. France.

(2) Laboratoire de Géologie Marine et Sémimentologie Appliquée. Centre Universitaire de Marseille-Luminy, Case 902, 13288 MARSEILLE Cédex. France.

(3) Laboratoire de Palynologie, Muséum National d'Histoire Naturelle, 62, rue de Buffon, 75005 PARIS. France.

RESUME

En vue de reconstituer l'évolution climatologique de la Méditerranée
Orientale au cours de la dernière déglaciation, l'étude des dépôts marins
profonds de ce bassin a été entreprise par deux approches complémentaires
des populations de foraminifères planctoniques : analyse des variations des
concentrations isotopiques de l'oxygène, analyse des variations des assemblages et application des fonctions de transfert.

Nos résultats montrent que la Méditerranée n'est pas un bassin exceptionnel.
Comme ceux des autres océans, ses sédiments profonds ont enregistré, avec
semble-t-il une grande précision, les principales fluctuations climatiques
de notre globe. La déglaciation (20.000 BP à l'Actuel) ne s'est pas effectuée de façon graduelle mais selon une suite de stades et d'interstades
glaciaires de courte durée, de l'ordre du millénaire. Ces stades ne corres-
pondent donc pas à des facteurs de nature astronomique mais à des phénomènes
locaux, propres à notre planète.

INTRODUCTION

L'objectif de notre étude était de reconstituer l'évolution des températures de surface de la Méditerranée Orientale au cours des derniesr 30.000 a,s (période couvrant la dernière glaciation-déglaciation), en analysant les dépôts marins profonds de cette mer et de comparer cette évolution à celles des océans et continents environnants.

D'après l'hypothèse de Broeker et van Donk (1970), les variations des compositions isotopiques de l'oxygène dans les océans représentent, depuis les derniers 3 Ma, les variations des volumes de glace stockés dans les calottes glaciaires et reflètent, en gros, les variations climatiques. Cette hypothèse est aujourd'hui largement confirmée pour les grands océans (Shackleton et Opdyke, 1973, Duplessy et al., 1981, etc.).

Toutefois dans la Méditerranée Orientale, bassin quasi-fermé et dont les communications avec l'Océan Mondial ont pu être entravées par les variations du niveau marin liées aux glaciations, un facteur local lié aux variations du rapport précipitations/évaporation a pu contribuer aux variations isotopiques des eaux du bassin et, dans une certaine mesure, masquer l'évolution climatique.

Nous avons utilisé, pour reconstituer cette évolution climatique, une approche pluridisciplinaire en appliquant simultanément trois techniques complémentaires pour l'étude des foraminifères planctoniques et benthiques de chaque colonne stratigraphique examinée, datée au ^{14}C.

1º - Les compositions isotopiques de l'oxygène et du carbone

2º - La nature des assemblages de faunes : chaudes, froides, etc.

3º - L'application des fonctions de transfert aux assemblages de faunes pour restituer les paléotempératures.

METHODOLOGIE

Pour notre travail, nous disposions des 32 carottes de la Campagne Océano-
graphique MEDOR-75 du N/O LE SUROIT. Après une étude exploratoire nous en
avons retenu sept, disposées d'est en ouest le long du bassin (fig. 1).

Les datations au radiocarbone ont été réalisées sur la fraction carbonatée
des sédiments, avec 5568 ans comme demi-période de vie du ^{14}C et sans cor-
rections de l'âge conventionnel obtenu (annulation réciproque de l'enrichis-
sement en ^{13}C et de l'âge apparent de l'eau de mer).

Les analyses isotopiques de l'oxygène et du carbone ont été effectuées selon
la technologie classique (Shackleton et al. 1973) sur un spectromètre de
masse VG 602 C.

Les analyses et comptages des microfaunes ont porté, pour chaque niveau, sur
300 à 500 individus après tamisage sur 150 mµ. L'indice faunistique climatique utilisé est la différence entre les sommes des espèces chaudes et froi-
fes. Nous avons choisi, d'après Blanc et al. (1976) comme espèces froides :
Globorotalia scitula, Globigerina quinqueloba, Globigerinita glutinata,
Globigerina pachyderma, G. clarkei, et comme espèces chaudes : *Globigerinoides*
ruber, G. trilobus-sacculifer, Globigerinella siphonifera, Orbulina universa,
Globigerina calida, G. rubescens, G. digitata. L'espèce tempérée-chaude
Globorotalia truncatulinoides a été réunie à ce groupe. Par contre
Globorotalia inflata, espèce vraiment transitionnelle, n'a pas été retenue
en raison de sa position trop intermédiaire.

Pour estimer les paléotempératures des eaux de surface à partir des assemblages de foraminifères, nous avons utilisé les fonctions de transfert
d'Imbrie et Kipp (1971) avec la matrice de référence mixte (espèces méditerranéennes et atlantiques) de Thunnell (1978). Toutefois dans certains assemblages des espèces dépassent en pourcentage leur représentation maximale
dans la matrice de comparaison et conduisent à des résultats aberrants.
Nous les avons, soit éliminés, soit, lorsqu'ils représentaient un intérêt
particulier, gardés mais signalés par un astéristique (fig. 2). Nous tra-
vaillons actuellement à améliorer cette matrice de comparaison.

DISCUSSION

Le problème majeur dans l'étude des dépôts marins, et spécialement dans les recherches de paléoclimatologie, est celui de leur remaniement possible par des processus mécaniques ou biologiques. En particuler la biotrubation affecte la plupart des sédiments océaniques, "homogénéisant" sur une dizaine de centimètres la couche superficielle.

Dans ce travail, peut-être à une exception près (Ks 52), toutes les carottes que nous avons étudiées semblent avoir été remaniées à des degrés divers. Nous rejoignons ainsi D. Stanley (1981) qui pense qu'il s'agit d'une règle générale dans la Méditerranée Orientale. Toutefois parmi les carottes les moins perturbées, deux, Ks 50 et Ks 76, présentent des courbes isotopiques $\delta^{18}O$ proches de celles publiées pour l'Atlantique (Duplessy et al., 1981) et peuvent, elles aussi, être comparées aux carottes des autres océans.

INTERPRETATION DES ANALYSES ISOTOPIQUES

Variations des compositions isotopiques des foraminifères benthiques

Les renseignements sur les eaux profondes sont apportés par les Fora-minifères benthiques. Les eaux de Méditerranée occidentale sont relative-ment peu oxygénées sur le fond et pauvres en sels minéraux. Les Forami-nifères benthiques sont donc très rares; *Uvigerina spp.* est généralement absente au-dessous de 1500 m environ. Cette espèce a cependant pu être analysée dans deux carottes : Ks 26 et Ks 50, dans les niveaux d'âge gla-ciaire ($\leq$ 18 000 ans). Les valeurs maximales de $\delta^{18}O$ sont comprises entre 4,75 et 4,80 ‰. Les formes récentes de cette espèce ont un $\delta^{18}O$ proche de 2,1 ‰. La variation de 4,8 - 2,1 = 2,7 ‰ enregistrée entre glaciaire et post-glaciaire correspond donc à la valeur minimale de l'effet glaciaire (stockage des glaces) en Méditerranée Orientale. C'est aussi l'effet enre-gistré par *G. bulloides* et d'autres Foraminifères benthiques dans le bassin occidental. Cette variation serait donc celle des calcites d'eau profondes et/ou d'eaux hivernales. Cette variation de 2,7 ‰ est supérieure à 1 ‰ à celle reconnue pour les calcites atlantiques pendant la même période. Cet effet supplémentaire de 1 ‰ serait le résultat d'un effet température et/ou d'évaporation/précipitation. Il implique, en tout cas, que la baisse de température des eaux hivernales ou des eaux profondes n'a pu dépasser 4° C pour le dernier glaciaire, dans tout le bassin.

Compositions isotopiques des foraminifères planctoniques

Les variations des paléoparamètres physico-chimiques de surface nous sont
données par l'analyse de *G. ruber*. Actuellement, cette espèce domine pendant
l'été et l'automne en Méditerranée orientale. Cette saison correspond à
celle des pluies tropicales sur l'Afrique. Mais rien ne dit que la saison
optimale n'a pas varié dans le temps et que les périodes glaciaires n'aient
pu favoriser la sélection d'écotypes hivernaux, aux caractéristiques mor-
phologiques et isotopiques différentes.

Pour l'ensemble des carottes la variation globale de $\delta^{18}O$ enregistrée par
G. ruber au cours de la dernière déglaciation varie entre 4 et 5,25 %₀ entre
maximum glaciaire et minimum inter-glaciaire (correspondant à la phase
Atlantique). Toutefois la variation "glaciaire/période actuelle" est plus
faible, et si l'on admet que la valeur actuelle de $\delta^{18}O$ de *G. ruber*
en Méditerranée Orientale est voisine de 0 %₀, on trouve pour les carottes
étudiées, entre le maximum glaciaire et l'actuel, un $\Delta\delta^{18}O$ compris entre
3,25 et 3,75 %₀. La valeur de 3,75 %₀ est supérieure à 1 %₀ à la variation
enregistrée par les Foraminifères benthiques. Cet effet peut résulter de
modifications locales dans le bilan évaporation/précipitation ou de changement dans la saison optimale de l'espèce. Si cet écart devait être imputé
uniquement à un effet de salinité dans les eaux superficielles, on peut
calculer que cet effet n'a pu dépasser la valeur donnée par la relation
$\Delta\delta^{18}O$ %₀ = 0,27 ΔS %₀, ce qui permet de calculer ΔS %₀ $\leq$ 3,7 %₀, valeur
compatible avec l'apport d'importantes masses d'eau douce par le Nil.

STRATIGRAPHIE ISOTOPIQUE ET GEOCHRONOLOGIE DE LA DERNIERE GLACIATION

Si l'on admet qu'en Méditerranée Orientale, comme dans les autres océans,
les courbes isotopiques $\delta^{18}O$ des Foraminifères de surface (ici *G. ruber*)
représentant essentiellement l'effet de stockage glaciaire amplifié par un
effet d'évaporation/précipitation, ces courbes peuvent être utilisées pour
une stratigraphie isotopique et être interprétées, en gros, en termes de
variations climatiques.

Pour les carottes peu perturbées, Ks 50 et 76, les courbes $\delta^{18}O$ présentent
un bon parallélisme avec les courbes publiées pour l'Atlantique (Duplessy et
al., 1981) et il est possible d'y reconnaître les terminaisons Ia et Ib,

approximativement aux mêmes dates. C'est une première indication que les
courbes isotopiques de ce bassin traduisent une évolution climatique sembla-
ble à celle des autres océans.

Chronologie glaciaire de la carotte Ks 52

Parmi les carottes généralement remaniées de cette étude, la carotte Ks 52
semble être une exception. Elle a été prélevée dans la fosse de Strabon
au sud de la Crète, fosse isolée des apports terrigènes, des courants de
fond et surtout caractérisée par des conditions anoxiques au cours de la
période couvrant la dernière déglaciation. L'absence de vie et d'organismes fouisseurs sur le fond a empêché toute bioturbation des sédiments, per-
mettant ainsi un enregistrement pratiquement non-perturbé de l'évolution
paléoclimatique du bassin pendant cette période.

Si nous admettons, comme hypothèse de travail, que toutes les couches de
Ks 52 n'ont subi aucune perturbation, ce qui peut ne pas être exact à tous
les niveaux, nous pouvons essayer d'interprêter comme variations climatiques
les données conjuguées des variations de composition isotopiques $\delta^{18}O$, des
microfaunes et des fonctions de transfert.

Les datations au radiocarbone : les causes du déluge

Le dépôt de sédiments non bioturbés permet une datation fine des niveaux de
la carotte Ks 52. Une des premières conséquences est la datation précise
de la dernière période de confinement du bassin, période qui correspond au
dépôt du sapropèle S1. Dans le Ks 52, ce sapropèle est daté 11.760 $\pm$
142 BP à 7.950 $\pm$ BP. Les datations de deux autres carottes, Ks 5 :
11.100 à 7.335 BP et Ks 76 : 11.200 à 8.200 BP, sont proches de ces chiffres.

Ces résultats montrent que, contrairement à un modèle largement accepté
(Olausson, 1961, Ryan, 1972, etc.), la dernière période -e confinement de la
Méditerranée Orientale ne correspond pas à l'arrivée des eaux de fonte de la
calotte nordique de glace eurasiatique. En effet, à partir de 13.000 BP,
ces eaux de fonte ont été déviées vers la mer de Norvège et n'alimentaient
pratiquement plus la Méditérranée (Grosswald, 1980). Par contre nos dates

du sapropèle S1 coïncident avec la période d'intensification de la mousson
sur l'Afrique équatoriale, sèche jusqu'alors et au déversement, par la voie
du Nil, d'importantes masses d'eaux douces dans la Méditerranée Orientale.
Ce phénomène a dû être particulièrement brutal et important puisque l'huma-
nité en a gardé la mémoire : le déluge, dont les dates rapportées par Platon,
coïncident avec l'établissement de la stagnation dans le Bassin Oriental.
De plus, cette intensification brutale de la mousson a intéressé toute la
ceinture équatoriale de la planète. Aussi nous pouvons proposer, comme hypothèse de travail, que les divers sapropèles du Bassin Oriental (S1 à S9,
etc.) correspondraient, eux aussi, à des changements climatiques de même
nature, une augmentation brutale de la pluviosité sur les zones équatoriales
du globe. D'après la courbe de Milankovich, ce phénomène semble correspondre
aux périodes de forte insolation de l'hémisphère Nord. Comme nous savons que
les interactions océan-atmosphère au-dessus de la zone tropicale sont un des
éléments fondamentaux de la circulation atmosphérique sur l'hémisphère Nord,
l'étude de ces anciens sapropèles pourrait contribuer à la reconstitution
des paléoclimats de cet hémisphère et plus particulièrement de l'Europe.
Enfin, on pourrait étendre cette hypothèse aux grandes périodes de stagnation
reconnues dans des anciennes mers des séries géologiques, par exemple dans
le Crétacé (Rossignol-Strick et al., 1982).

Paléoclimatologie de la carotte Ks 52

Les diverses courbes de la carotte Ks 52, compositions isotopiques de l'oxygène de *G. ruber* (correspondant aux variations du volume des glaces continentales), indice des microfaunes et températures des eaux superficielles données par les fonctions de transfert, présentent entre elles une bonne concordance (fig. 2). Dans le détail toutefois cette concordance n'est pas toujours parfaites et, à quelques niveaux, on observe des résultats contradic-
toires.

Ces courbes montrent qu'entre le maximum de la dernière glaciation et nos
jours la température des eaux superficielles de ce site de la Méditerranée
Orientale n'a pas augmenté de façon régulière mais selon un dessin en dents
de scie correspondant à des fluctuations du climat. Nous sommes ainsi con-
duits a comparer l'évolution paléoclimatique de ce bassin à celle des continents et océans proches ou même lointains.

Une telle comparaison n'est pas aisée, car de nombreux désaccords subsistent
entre les dates proposées par différents auteurs. Toutefois, un schéma géné-
ral et des courbes synthétiques de géochronologie globale (Kind 1972, Dreimanis
et Karrow 1972, etc.) se dégagent des très nombreuses études ponctuelles.
Pour notre comparaison, nous nous baserons sur les travaux des auteurs sui-
vants : Evin 1979, pour l'Europe du Sud; Mörner 1971 et 1972, Mangerud et
al. 1974, Mangerud 1980, pour la Scandinavie; Newstad 1971, Raukas et al.,
1972, pour la plate-forme Russe; Hammer et al., 1978, pour le Groënland;
Dreimanis et Karrow 1972, Thorson 1980, pour l'Amérique du Nord; Kind 1972,
pour la Sibérie, etc., ainsi que sur la série des "International Geological
Correlation Programme" Project 73/1/24 de l'UNESCO-IUGS.

L'examen de la figure 2 montre une bonne corrélation entre les âges extra-
polés de la séquence de pics froids et chads du site 52 et les âges des
stades et interstades glaciaires définis, à terre, sur les divers continents
de notre planète (tableau 1). On observe, dans Ks 52, un maximum glaciaire
vers 20.000 BP. Le premier réchauffement, qui commence vers 16.000 BP, correspondrait à l'interstade Lac Erié. Il est suivi, vers 15.000 BP, d'un
important refroidissement qui correspondrait au stade Cary. L'interstade
Mackinaw est assez mal défini sur la courbe $\delta^{18}O$, mais les faunes chaudes
abondantes à partir de 14.000 BP, pourraient lui correspondre. Le stade
froid Port Huron, reconnaissable à partir de 13.340 BP, sur la courbe $\delta^{18}O$,
présente une alternance de niveaux à faune très froide et des niveaux plus
chauds. Au-dessus, nous observons avec des âges bien correlés, la série
européenne classique, Bolling, Dryas moyen (Older Dryas), Allröd, Dryas
recent, Préboréal et Atlantique.

Les courbes de la figure 2 montrent pour la première fois que, dans les
océans aussi, la séquence de déglaciation est, dans le détail, bien plus
compliquée que l'étude de carottes océaniques affectées par la bioturbation
ne permettrait de la prévoir (Ruddiman et McIntyre 1973, 1981, Duplessy et
al., 1980, 1981, etc.). En particulier la période appelée terminaison I
(Broecker et van Donk, 1970) comprendrait bien tous les stades et interstades
glaciaires définis sur les continents, mais oblitérés par la bioturbation
dans la plupart des dépôts océaniques. Toutefois les épisodes les plus im-
portants ressortent du bruit de fond de la courbe $\delta^{18}O$. Ainsi les premières
phases du réchauffement, tous stades confondus (Lac Erié au Bolling), ont
été appelées terminaison Ia, datée vers 14.500 BP, et le brutal passage
Dryas/Recent/Préboréal, daté 10.000 BP, terminaison Ib (Duplessy et al. 1981).

Malgré cette bonne correspondance chronologique des événements de Ks 52, nous
devons constater que certaines valeurs des variations $\delta^{18}O$ de cette carotte
ne peuvent être expliquées par les équations paléoclimatologiques classiques.
Si la variation de $\delta^{18}O$ entre le maximum glaciaire et l'actuel est la même
que celle enregistrée par *G. Bulloïdes* en Atlantique (Duplessy et al. 1981)
ainsi que la variation "maximum glaciaire - Dryas récent", 1 ‰ entre 43,5
et 24,5 cm en Ka 52, il n'en va pas de même pour les fluctuations de courte
durée et les grands pics négatifs. Ainsi entre le maximum glaciaire et la
terminaison Ia, vers 12.000 BP, on observe une variation de 4,5 ‰ en Ks 52
(42.5 et 28 cm) contre 1,3 ‰ en Atlantique et, pour la terminaison Ib vers
8.000 ans, 3,9 ‰ en Ks 52 (24,5 et 17,5 cm) contre 1,5 ‰ en Atlantique.
Cette forte baisse des $\delta^{18}O$ peut résulter de plusieurs phénomènes : un réchauffement plus important des eaux superficielles en Méditerranée (mais ce
n'est pas confirmé par les fonctions de transfert), un effet de salinité qui
se superpose à l'effet glaciaire avec dilution des eaux superficielles
(Rossignol et al. 1981) et enfin la possibilité de remaniements non discer-
nables à l'examen optique, à certains niveaux, de faunes plus anciennes.

CONCLUSIONS

Les données conjuguées de deux approches, analyse des variations isotopiques
de l'oxygène et analyse des variations des assemblages de microfaune montrent
que la Méditerranée Orientale n'est pas un bassin exceptionnel. Comme ceux
des autres océans, ses sédiments profonds ont enregistré les grandes fluctuations climatiques de notre planète. La déglaciation (20.000 BP à l'Actuel)
ne s'est pas effectuée de façon graduelle mais selon une suite de stades
froids et d'interstades plus chauds ou chauds. La durée de chaque épisode
climatique est assez courte, quelques siècles au millénaire. Ils ne corres-
pondent donc pas à des facteurs de nature astronomique, comme des variations
de l'orbite terrestre, mais à des phénomènes plus locaux, par exemple des
variations des conditions de notre planète.

De plus, si l'on admet que la carotte Ks 52 n'est pas perturbée, ce site a
enregistré, avec un grand détail et une bonne correspondance chronologique,
tous les stades et interstades glaciaires définis sur les continents et pourrait présenter une coupe-type du Bassin Oriental.

BIBLIOGRAPHIE

Blanc, F. et al. (1976). - *Paleogeogr. Paleoclim. Paleoecol.*, *20, 277-296.*

Broecker, W.S. et al. (1970). - *Rev. Geogh. Space Phys.*, *8, 1, 169-198.*

Dreimanis, A. et al. (1972). - *24th Int. Geol. Congr.*, *Montréal, Section 12,*
5-15.

Duplessy, J.C. et al. (1981). - *Paleogeogr., Paleoclim., Paleoecol., 35,*
121-144.

Evin, J. (1979). - *Coll. Int. CNRS n° 271, 5-13.*

Grosswald, M.G. (1980). - *Quat. Res., 13, 1-32.*

Hammer, C.U. et al. (1978). - *J. Glaciology, 20, 82, 3-26.*

Imbrie, J. et al. (1971). - In : K.K. Turekian Ed. *Late Cenozoix Glacial*
Ages. Yale Un. Press. 71-181.

IUGS-UNESCO. International Geological Correlation Programme.
Project 73/1/24. Reports 1 to 7 (1982).

Kind, N.Y. (1972). - *24th Int. Geol. Congr.*, *Montréal, Section 12, 55-61.*

Mangerud, J. (1980). - Studies in Late Glacial N.W. Europe.
J.J. Lowe et al. Ed. Pergamon Press. 23-30.

Mangerud, J. et al. (1974). - *Boreas, 3, 109-128.*

Mörner, N.A. (1971). - *Geol. Fören. Stockolm. Förhandl., 93, 236-238.*

Mörner, N.A. (1972). - *24th Int. Geol. Congr.*, *Montréal, Section 12,*
72-79.

Newstad, M.I. (1971). - *Geolg. Fören. Stockolm Förhand., 93, 103-115.*

Olausson, E. (1961). - *Swedish Deep-Sea Exp. Repts., 1947-1948, 8,*
337-391.

Raukas, A.V. et al (1972). - *24th Int. Geol. Congr.*, *Montréal, Section 12,*
97-102.

Rossignol-Strick, M. et al (1982). - *Nature, 295, 5845, 105-110.*

Ryan, W.B.F. (1972). - The Mediterranean Sea.
Dowden, Hutchinson and Ross, Stroudsburg, 149-169.

Shakleton, N.J. et al. (1973). - *Quat. Res., 3, 39-55.*

Stanley, D. (1981). - *Geo-Marine Letters, 1, 77-83.*

Thorson, R.M. (1980). - *Quat. Res., 13, 303-321.*

Thunnel, R.C. (1978). - *Marine Micropal., 147-173.*

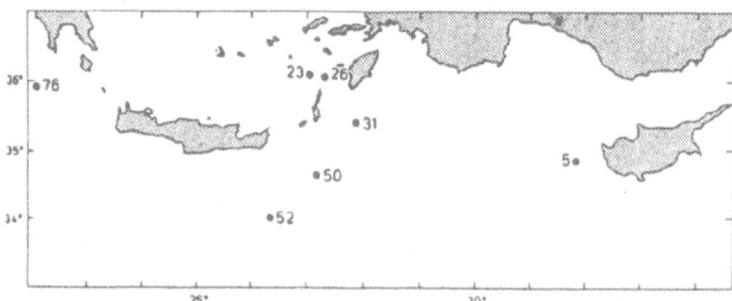

Figure 1 : Position des carottes étudiées.

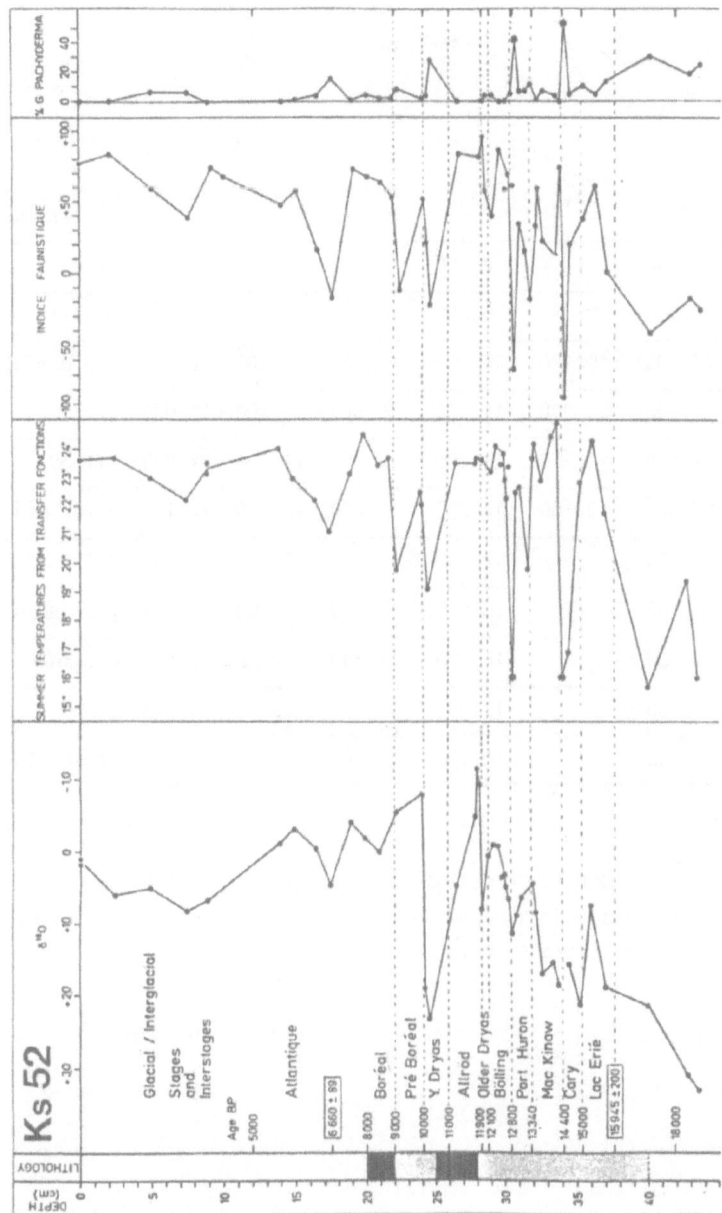

Figure 2 : Carotte Médor 75-Ks 52. Stades et interstades du maximum de la
dernière glaciation à l'actuel définis par les courbes $\delta^{18}O$, les
paléotempératures données par les fonctions de transfert et les
indices des microfaunes. Les datations au ^{14}C sont encadrées
toutes les autres dates sont interpolées.

Tableau 1

Sibérie	Amérique du Nord	Camp Century	Scandi-navie	Plate-forme Russe	France Europe Sud	Ks 52	Stades et interstades
7.900	7.000		8.000		8.000		Atlantique
8.300	8.500		9.000	9.000	9.000	9.000	Boréal
10.300	10.000	10.100	10.000	10.200	10.000	10.000	Pré Boréal
11.400	11.000	10.900	11.000	10.800	10.800	11.000	Dryas récent
11.800	11.800	11.850	11.800	12.000	11.800	11.900	Allröd
12.200	12.000	12.100	12.000	12.200	12.400	12.100	Dryas moyen
13.000	12.700		12.500	12.700	13.000	12.800	Bolling
	13.300		13.500	13.200	13.600 ?	13.340	Port Huron
	14.000		14.000	13.700	14.400 ?	14.400	Mackinau
15.000	15.000		15.000	15.500	15.000	15.000	Cary
16.000	16.000		16.000	16.000	16.000	16.000	Lac Erié

Datations publiées pour le début des stades et interstades de la dernière glaciation-déglaciation dans le monde, comparées aux âges interpolés pour la carotte Ks 52 de Méditerranée Orientale.

LATE-GLACIAL CLIMATE HISTORY FROM ICE CORES [*]

H. Oeschger, J. Beer, U. Siegenthaler and B. Stauffer

Physics Institute, University of Bern, Switzerland

W. Dansgaard

Geophysical Isotope Laboratory, University of Copenhagen, Denmark

C.C. Langway

Dept. of Geological Sciences, State University

of New York at Buffalo, Amherst, N.Y., USA

ABSTRACT

Ice cores contain information on climatic variations and their causes. Recent results obtained on the new deep ice core drilled in 1981 at Dye 3, South Greenland, in the frame of the US-Danish-Swiss Greenland Ice Sheet Program are :

- Comparison of the $\delta\,^{18}O$ variations in the Greenland ice cores with those in European lake carbonate exhibits strong similarities and provides time marks (13,000, 11,000, 10,000 B.P.) for the Late-Glacial section of the ice cores;

- CO_2 concentration measurements in the occluded air indicate low (180-200 ppm) CO_2 concentrations 30,000 to 15,000 B.P. and an increase to ca. 300 ppm around 13,000 B.P.. The CO_2 increase might reflect a change in the ocean circulation at the end of the last glaciation and could have contributed to the establishment of the Holocene environmental conditions;

- ^{10}Be concentration measurements on samples covering the last 50,000 yaers show a correlation with $\delta\,^{18}O$, low $\delta\,^{18}O$ values corresponding to high ^{10}Be concentrations (atoms per g of ice). Probably this mainly reflects changes in the rate of precipitation in the northern hemisphere.

Based on the ice core information climatic events during the Glacial-Postglacial transition are discussed.

[*] Condensed and slightly modified version of a paper presented at the Ewing Symposium, Fall 1982, at Lamont-Doherty Geophysical Observatory, to appear in the proceedings of this symposium.

1. Introduction

In the last 5 to 10 years the potential of ice core studies as a means
for reconstructing the history of the environment system and for better understanding its mechanisms has been recognized by a broad spectrum
of disciplines involved in earth and planetary sciences. Reasons of this
development are :

- the succesful ice core drilling in polar areas, e.g. the drilling through
  the Greenland ice sheet in 1981 at Dye 3, South Greenland, in the frame
  of the US-Danish-Swiss Greenland Ice Sheet Program (GISP) [AGU, 1983];

- recent progress in the development of highly sensitive field and laboratory analytical techniques. Examples are spectrocopic CO_2 analysis, accelerator mass spectrometry, electrical conductivity measurements and ion
  chromatography, enabling the study of new types of environmental parameters;

- strengthened attempts to understand environmental system processes in
  view of possible natural and/or anthropogenic climatic changes.

In this paper we discuss information obtained from ice core analyses covering about the last 60,000 years of the Dye 3 ice core. First, the $\delta^{18}O$
records of the Greenland ice core and of 14_C dated lake sediments from
Central Europe are compared. Strong similarities of the late glacial climatic events in the two regions are revealed which permits to define time
marks for the Greenland $\delta^{18}O$ profiles. Then CO_2 concentrations af the air
trapped in the ice are presented and the implications regarding carbon cycle and climate are discussed. Furthermore 10_{Be} data are given together
with possible explanations for the observed changes. Based on the information on these three parameters conclusions regarding the climatic evolution during the Glacial-Postglacial transition are drawn.

2. The climate system information from ice cores

To obtain an impression of the climate information of ice cores we consider the global average energy balance on the top of the atmosphere :

$$S\pi R^2(1-A) = 4\pi R^2 \sigma T_S^4 (1-B)$$

With   S = solar constant

       R = radius of earth

       A = alvedo (reflected fraction of solar irradiation)

       σ = Stefan-Boltzmann constant

       B = fraction of infrared radiation (emitted from surface)
             absorbed in atmosphere and reemitted back to surface

       T_S = surface temperature

Variations of the parameters S, A and B lead to changes in the passive parameter T_S, the earth's surface temperature. Information on all these four
parameters is recorded in ice cores :

Variations in the emission of solar plasma lead to variations in the magnetic shielding of the inner part of the solar system against galactic cosmic radiation. Fluctuations of the cosmic ray flux reaching the earth are reflected e.g. in variations of the production of radioactive nuclei in the earth's atmosphere. These production rate changes are recorded in tree-rings as $^{14}C/^{12}C$ ratio changes and in changes of the cosmogenic radioisotopes (e.g. ^{10}Be, ^{36}Cl) contents of precipitation (chapter 6). It seems plausible that changes of other solar properties like luminosity and ultraviolet emission are somehow related to solar plasma emission changes.

Changes in the ^{10}Be content of ice cores thus may reflect changes in solar parameters (like S) which could influence climate.

Variations in atmospheric turbidity and thus in albedo (A), are induced by vulcanic eruptions leading to stratospheric dust layers. Solid electrical conductivity measurements on ice cores enable the identification of volcanic dust and therefore contribue to the reconstruction of the history of atmospheric turbidity (Hammer et al., 1980). The solid electrical conductivity measurements are supported by chemical measurements, e.g. contents of anions.

Air bubles in the ice constitue samples of the ancient atmosphere. Measurements of the gas composition reveal variations of the contents of infrared active gases like CO_2 (chapter 4) and CH_4 (Craig, 1982). Ice seems to be the only natural archive in which essentially undisturbed air samples are stored.

Finally changes in temperature (T_S) are revealed as changes in the $^{18}O/^{16}O$ ratio of the water molecules (chapter 3). Water little depleted in ^{18}O relative to sea water reflects warm, water strongly depleted in ^{18}O, cold climate. In favorable cases the high resolution of the ice core information enables reconstruction of seasonal ^{18}O variations up to 10,000 y back in time. In addition air temperature excursions above $0°C$ lead to melting (Herron et al., 1981) and the melt layer fraction can be used as an indicator of the lenght and intensity of the summer melting periods. At locations on ice caps where the mean annual temperature is $\leq -30°C$ melting of ice is negligible.

For the understanding of characteristic properties of the complex climate system studies of its response to external forcing in the past is especially promising. The core parameters provide information both on the external forcing, (S, A and B) and the system's response T_S as will be discussed in the next chapters.

3. Comparison of the $\delta^{18}O$ records from Dye 3 and from Central European lake sediments

A continious $\delta^{18}O$ profile has been measured along the newly recovered deep ice core from South Greenland (Fig. 1). Dansgaard et al. (1982) compared the features of this $\delta^{18}O$ profile with those observed for the Camp Century ice core drilled in the 1960ies. It appears that for the late Wisconsin essentially all the oscillations in the Dye 3 record can be correlated with similar features in the Camp Century record.

$\delta^{18}O$ variations of precipitation are recorded not only in polar ice but also in lake carbonate sediments of authigenic origin. Late Glacial $\delta^{18}O$

profiles measured on European lake sediments show strong variations which can be correlated with climatic changes as inferred from pollen analysis. Fig. 2 shows as an example the $\delta^{18}O$ record of a sediment profile form Lake Gerzensee, Switzerland, (Eicher, 1980). At several other sites in Central Europe $\delta^{18}O$ profiles could be obtained which exhibit the same major features (Eicher and Siegenthaler, 1976; Eicher et al., 1981). The climatic history involves a first major warming around 13,000 B.P. (^{14}C age) accompagnied by a rapid increase in $\delta^{18}O$, corresponding to the Bølling-Allerød warm phase (pollen zones Ib-II); a transition to the Younger Dryas cold phase (pollen zone III) at about 11,000 B.P., and the final transition to the Holocene at about 10,000 B.P. The three climatic changes are marked by abrupt $\delta^{18}O$ shifts which indicate that the climate in Central and NW Europe changed drastically at those times.

In Fig. 3 that part of the Dye 3 $\delta^{18}O$ profile which represents the Gliacial-Postglacial transition period is compared with the Gerzensee $\delta^{18}O$ profile. The similarity of the $\delta^{18}O$ fluctuations between 1785 m and 1815 m depth with those observed in the lake sediments is striking : not only the major isotopic shifts but also minor fluctuations can be correlated. This strongly suggests that the two $\delta^{18}O$ profiles reflect the same evolution of climatic events. We therefore assume for the Greenland $\delta^{18}O$ record the approximate time marks on the ^{14}C time scale of 13,000 B.P., 11,000 B.P. and 10,000 B.P., as indicated in Fig. 2.

4. CO_2 concentrations along the Dye 3 ice core

The second parameter to be discussed is the CO_2 content of the air occluded in the Dye 3 ice. During the transition from firn to ice, at Dye 3 presently ocurring in a depth interval from 65 to 80 m, air parcels are occluded as bubbles in the freshly formed ice. Thus ice continuously collects air samples. As has been shown by different groups (Raynaud and Delmas, 1977; Scholander et al., 1961; Stauffer, 1981) the concentration of the major constituents of the occluded air (N_2, O_2, Ar) closely reflects the atmospheric composition if summer melting can be excluded. We therefore assume that also the CO_2 concentration of this air corresponds to the original atmospheric value as long as no melt layers are present. If melting occurs, higher CO_2 concentrations are observed because CO_2 is strongly soluble in liquid water and during refreezing gets trapped preferentially compared to the main air gases. The air in the bubbles represents the composition of the atmosphere at the time of air occlusion; it is therefore younger than the surrounding ice.

In Fig. 4 the CO_2 concentrations observed in the Dye 3 core between 1780 and 1940 m depth, corresponding approximately to the period 9000 to 60,000 B.P., are represented, together with $\delta^{18}O$ and ^{10}Be data. CO_2 concentrations were determined by crushing ice samples of ca. 1 cm^3 and measuring the CO_2 content of the escaping air by means of an infrared laser spectrometer. The precision of the analyses is about $\pm$ 6 ppm. It can be assumed that during the glacial period no summer melting occurred so that the CO_2 composition of the ice age atmosphere is reflected without alterations. This does not hold for parts of the Holocene ice which are affected by melting (Stauffer et al., 1983).

The following main results are observed (Fig. 4). For the late Wisconsin (1829 m to 1870 m, corresponding to about 14,000 B.P. to 30,000 B.P.) low CO_2 valies in the range of 180 to 200 ppm are obtained, in agreement with earlier measurements on the ice cores from Camp Century,

Greenland, and Byrd Station, Antarctica (Neftel et al., 1982), as well as
on other Antarctic ice cores (Delmas et al., 1980). The CO_2 increase as
observed in the Dye 3 core occurs at almost the same depth as the $\delta^{18}O$
increase indicating the first warming at the end of the last glaciation,
ca. 13,000 B.P. The depth at which the CO_2 shift takes place can, based
on the present data, not be exactly defined; it is a few m below the
$\delta^{18}O$ shift. At Dye 3 the age differences between the trapped air and the
surrounding ice is at present ca. 100 years and was probably somewhat larger
13,000 B.P. because of smaller accumulation rate and lower temperature.
The annual layer thickness can be estimated from Fig. 3 to 1 cm or
slightly less, so that the depth differences between the shifts of $\delta^{18}O$
and CO_2 would be between 1 and 3 m if both changes actually occurred at
the same time. We therefore conclude that the two parameters changed
almost simultaneously with a possible time difference of a few centuries
at maximum.

For the period 13,000 to 11,000 B.P. relatively constant CO_2 values in a
band between 300 and 320 ppm are observed. These values may be slightly
affected by melt layer ice which is enriched in CO_2. At about 11,000
B.P., the beginning of the Younger Dryas cold period, two low CO_2 concentration values of about 250 ppm are observed, followed again by a value
slightly above 300 ppm and a series of increasing values from 280 to 330
ppm at the Younger Dryas-Preboreal transition.

Considering also the CO_2 data from Camp Century, Greenland, and Byrd
Station, Antarctica, we conclude that at the end of the last glaciation
the CO_2 concentration of air occluded in polar ice from Greenland as well
as from Antarctica increased from 180 - 220 ppm to 280 - 300 ppm. The
transitions are parallel to the $\delta^{18}O$ shifts.

4. Climatic implications of the atmospheric CO_2 increase during the Glacial-
   Postglacial transition

   Estimates of global temperature change based on different types of climate
   models indicate an almost logarithmic dependence of temperature on the
   atmospheric CO_2 concentration. The mean global temperature change corresponding to a CO_2 increase by a factor of 1.5 from 200 ppm to 300 ppm,
   as calculated by climate models, is about 1.5 ± 0.5^oC. Climate models
   also indicate a strong increased warming over ice-covered areas : The
   warming which probably occurred as a result of the global CO_2 increase
   therefore was probably amplified in regions of sea ice and on the continental ice shelves, leading to additional melting of continental ice and
   the disappearance of sea ice.

   A CO_2 induced warming effect at the end of the last glaciation would have
   provided part of the interhemispheric coupling so far missing in ice age
   theories since irradiation changes due to variations in the earth's
   orbital elements are not in phase in the two hemispheres. Such a strong
   interhemispheric coupling is, however, a necessity if the climatic change
   in both hemispheres occurred simultaneously. The rising CO_2 level in the
   end phase of the last glaciation influenced the radiation balance of the
   earth and amplified the general warming tendency. Besides that one might
   speculate that it contributed to the regrowth of vegetation due to CO_2
   fertilization, which involved an albedo decrease over the continents,
   thus providing an indirect CO_2-climate effect.

Taken together these effects resulting from a global CO_2 increase may have significantly contributed to the establishment of the present interglacial climatic conditions.

5. The ^{10}Be record in the Dye 3 ice core

The development of accelerator mass spectrometry rendered feasible the measurements of the number of ^{10}Be atoms in one kg of precipitation. ^{10}Be ($T_{\frac{1}{2}} = 1.5 \times 10^6$y) is mainly produced by spallation reactions on atmospheric nitrogen and oxygen. The freshly produced ^{10}Be becomes attached to aerosols within a short time and after a mean residence time in the atmosphere of ca. 1 to 2 years is deposited on the earth's surface by precipitation. ^{10}Be removed from the atmosphere is stratigraphically preserved in polar ice cores. Its concentration shows variations due to production rate changes and variations in atmospheric mixing and deposition. Recent ^{10}Be measurements on ice samples from Dye 3 covering the last seven sunspot cycles seem to reflect the known variation of cosmic ray interaction with the atmosphere over the 11 year suspot cycle (Beer et al., 1983).

From the Maunder minimum perdio (1640 to 1710) of the quiet sun Raisbeck et al., (1981) found ^{10}Be concentrations in Antarctic ice higher by a factor of 1.5 than observed for other periods.

More detailed measurements on ice samples from Dye 3 which we performend at the Zürich accelerator mass spectrometer have confirmed this result (Beer et al., in preparation). Besides production rate, other factors may influence ^{10}Be conconcentration. In periods of major climatic change, changes in the ^{10}Be concentration in precipitation may also have been caused by changes in the atmospheric mixing and circulation mode or by variations in precipitation rates.

The results from the Dye 3 core *) covering the last 30,000 years show a correlation both with δ^{18}O and CO_2 (Fig. 4) : low ^{10}Be concentrations are found in samples with high δ^{18}O and high CO_2 concentrations. Several explanations are possible. At present we favour the explanation that the variations of ^{10}Be concentration, expressed as atoms per g of water, mainly reflect changes of the precipitation rate. For the end of the Glaciation, ice age theories point at the possibility of reduced precipitation leading to a starvation of the ice sheets. The high ^{10}Be concentrations between 1820 and 1870 m depth might well reflect this phenomenon. Interestingly enough ^{10}Be and CO_2 correlate with δ^{18}O even for relatively short fluctuations, e.g. a low ^{10}Be concentration at 1885 m depth corresponds to high δ^{18}O values and a high ^{10}Be concentration at 1785 m depth corresponds to low ^{18}O (Younger Drays) values.

Though we consider precipitation rate changes as the major cause for the ^{10}Be fluctuations between 10,000 and 30,000 B.P., we do not exclude that they may partly be the result of changing solar and terrestrial magnetism, especially because solar-induced ^{10}Be variations by a factor of 1.5 to 2 have indeed been found for the Maunder minimum.

*) The ^{10}Be measurements were performed at the accelerator mass spectrometer of the Labor für Kernphysik of ETH, Zürich.

6. Conclusions

Measurements of $\delta^{18}O$, CO_2 and ^{10}Be on ice cores provide new insight into climatic processes at the glacial - postglacial transition. In the following we give the major conclusions and try to answer some questions which since a long time plan an important role in climate theories. Changes of the position of the oceanic polar front in the North Atlantic (Ruddiman and McIntyre, 1981), in the period 16,000 to 10,000 B.P. determined the climatic events in Europe and Greenland as reflected in the climatic parameters recorded in ice cores, lake sediments and peat bogs. Around 13,000 B.P. a first warming took place, expressed by the retreat of the deglacial polar front, a significant increase of $\delta^{18}O$ values in Greenland ice and in Central European lake sediments, combined with a shift of the vegetation from tundra to forest in Europe. This first warming was accompanied by a CO_2 increase which caused a significant change in the global radiation balance, equivalent to a global warming of ca. 1.50C. This phenomenon could have lead to a retreat of the sea ice cover in both hemisphere, providing a positive feedback to the developing climatic transition. In addition the CO_2 increase may have supported the vegetation regrouwth which also caused a reduction of the global albedo, another positive feedback. 11,000 B.P. a reestablishment of glacial conditions with a strong advance of the deglacial polar front occurred, possibly caused by a rapid disintegration of continental ice masses (Ruddiman and MasIntyre, 1981). This is reflected in the $\delta^{18}O$ profiles both in Greenland ice and in European lake sediments. Deep sea cores show a decrease of foraminiferal productivity and therefore missing supply of nutrient-rich deep water. Some ice core samples of this period also show low CO_2 concentrations. 10,000 B.P. the second (final) warming took place, as reflected by high $\delta^{18}O$ values, high CO_2 concentrations and relatively low ^{10}Be concentrations in the ice core and high foraminiferal activity, in the North Atlantic ocean, indicating a strong retreat of the deglacial polar front.

By means of more core drillings and further detailed CO_2 and $\delta^{18}O$ analyses, using the CO_2 increase as a time mark, it should become possible to check if the transition Glacial - Postglacial was synchronous in both hemispheres.

The atmospheric CO_2 concentreation is regulated by changes in the ocean's total CO_2 concentration, alkalinity, temperature and the content of nutrients (PO_4 and NO_3) that controls the biospheric activity in the surface ocean and thus the reduction of the CO_2 partial pressure relative to that of the average ocean water. Broecker (1982) has emphasized the role of phosphorous storage in the sediments as an important mechanism controlling the atmospheric CO_2 content to explain the fast CO_2 rise within a few centuries at about 13,000 B.P. This process is too slow to explain the observed rapid CO_2 change and we propose that CO_2 changes were a consequence of a change in the ocean's turnover rates, a slow turnover (ice age) leaving enough time for a complete nutrient consumption involving low pCO_2, and a faster turnover (Holocene) leading to incomplete nutrient consumption with a correspondingly higher pCO_2. It will be interesting to study by means of precise CO_2 measurements on ice cores if significant CO_2 variations also occurred during the Glacial and in the Holocene.

Acknowledgments

The authors express their gratitude to all those who contributed with great
motivation to these results : in the development and construction of drilling
and analytical equipment, drilling operation, analytical work in the field
and in the laboratory and in data interpretation.

Our efforts in ice core research have become possible thanks to the support
during many years by the U.S. and Swiss National Science Foundations, the
Danish Commission for Scientific Research in Greenland, the Danish Natural
Science Research Council and the U.S. Department of Energy.

References

AGU, Proceedings of the Spring meeting 1982, Philadephia, Union session :
    Greenland Ice Sheet Program, 1983, in press.

Beer, J., Andrée, M., Oeschger, H., Stauffer, B., Balzer, R., Bonani, G.,
    Stoller, Ch., Suter, M., Wölfli, W. and Finkel, R., Temporal ^{10}Be
    variations in ice, Radiocarbon, Vol., 25, No 2, 1983, pp. 269-278.

Berger, W.H., Johnson, R.F., Killigley, J.S., Unmixing of the deep-sea
    record and the deglacial meltwater spike, Nature, 269, 661-663. 1977.

Broecker, W.S., Ocean chemistry during glacial time, Geochim. Cosmochim. Acta,
    46, 1689-1705, 1982.

Craig, H., Chou, C.C., Methane : The Record in Polar Ice Cores, Geophysical
    Research Letters, 1982, in press.

Dansgaard, W., Clausen, H.B., Gundestrup, N., Hammer, C.U., Johnsen, S.F.,
    Kristinsdottier, P.M., Reeh, N., A new Greenland deep ice core,
    Science, 218, 1273-12-7, 1982.

Delmas, R.J., Ascencio, J.M., Legrand, M. Polar ice evidence that atmospheric
    CO_2 20,000 yr B.P. was 50 % of present, Nature, 284 (5752), 155-157,
    1980.

Duplessy, J.C., Delibrias, G., Turon, J.L., Pujol, C., Duprat, J., Deglacial
    warming of the northeastern North Atlantic Ocean : correlation with
    the palaeoclimatic evolution of the European continent, Palaeogeogr.,
    Palaeoclimat. Palaeoecol., 35, 121-144. 1981.

Eicher, U., Pollen- und Sauerstoffisotopenanalysen an spätglazialen Profilen
    vom Gerzensee, Faulenseemoos und vom Regenmoos ob Boltigen, Mitt.
    Naturforsch. Ges. Bern, 37, 65-80, 1980.

Eicher, U., Siegenthaler, U., Palynological and oxygenisotope investigations
    on Late-Glacial sediment cores from Swiss lakes, Boreas, 5, 109-117,
    1976.

Eicher, U., Siegenthaler, U., Wegmüller, S., Pollen and oxygen isotope ana-
    lyses on Late- and Post-Glacial sediments of the Tourbière de Chirens
    (Dauphiné, France), Quat. Res., 15, 160-170, 1981.

Hammer, C.U., Clausen, H.B., Dansgaard, W., Greenland ice sheet evidence of
         post-glacial volcanism and its climatic impact, Bature, Vol. 288,
         pp. 230-235.

Herron, M.M., Herron, S.L., Langway, C.C. Climatic Signal of ice melt
         features in southern Greenland, Nature, Vol. 293, pp. 389-391.

Imbrie, J., Imbrie, J.Z., Modeling the climatic response to orbital varia-
         tions, Science, 207, 943-953, 1980.

Neftel, A., Oeschger, H., Schwander, J., Stauffer, B., Zumbrunn, R., Ice core
         sample measurements give atmospheric CO_2 content during the past
         40,000 yr, Nature, 295, 220-223, 1982.

Raisbeck, G.M., Yiou, F., Fruneau, M., Loiseaux, J.M., Lieuvin, M., Ravel,
         J.C. and Lorius, C., [10]Be concentrations in Antarctic ice during past
         30,000 years, René Bernas report LRB 81-03, 1981.

Raynaud, D., Delmas, R., Composition des gaz contenus dans la glace polaire,
         Proceedings of the Int. Symp. on Isotopes and Impurities in Snow and
         Ice. IAHS-Publication No 118, pp. 377-81, 1977.

Ruddiman, W.F., McIntyre, A., The North Atlantic during the last deglaciation,
         Palaeoecol., 35, 145-214, 1981.

Scholander, P.F., Hemmingsen, E.A., Coachman, L.K., Nutt, D.C., Composition
         of gas bubbles in Greenland icebergs, Glaciology, 3, 813-822, 1961.

Siegenthaler, U., Oeschger, H., Heimann, M., ^{14}C variations caused by changes
         in the global carbon cycle, Radiocarbon, 22, 177-191, 1980.

Stauffer, B., Die Zusammensetzung der Luft in natürlichem Eis, Z.f.
         Gletscherkunde und Glazialgeologie, 17, 57-78, 1981.

Stauffer, B., Neftel, A., Oeschger, H., Schwander, J., CO_2 concentration in
         air extracted from Greenland ice samples, Proceedings AGU spring
         Meeting 1982, Philadelphia, 1983, in press.

Worthington, L.V., Genesis and evolution of water masses, Meteorol. Monographs,
         8, 63-67, 1968.

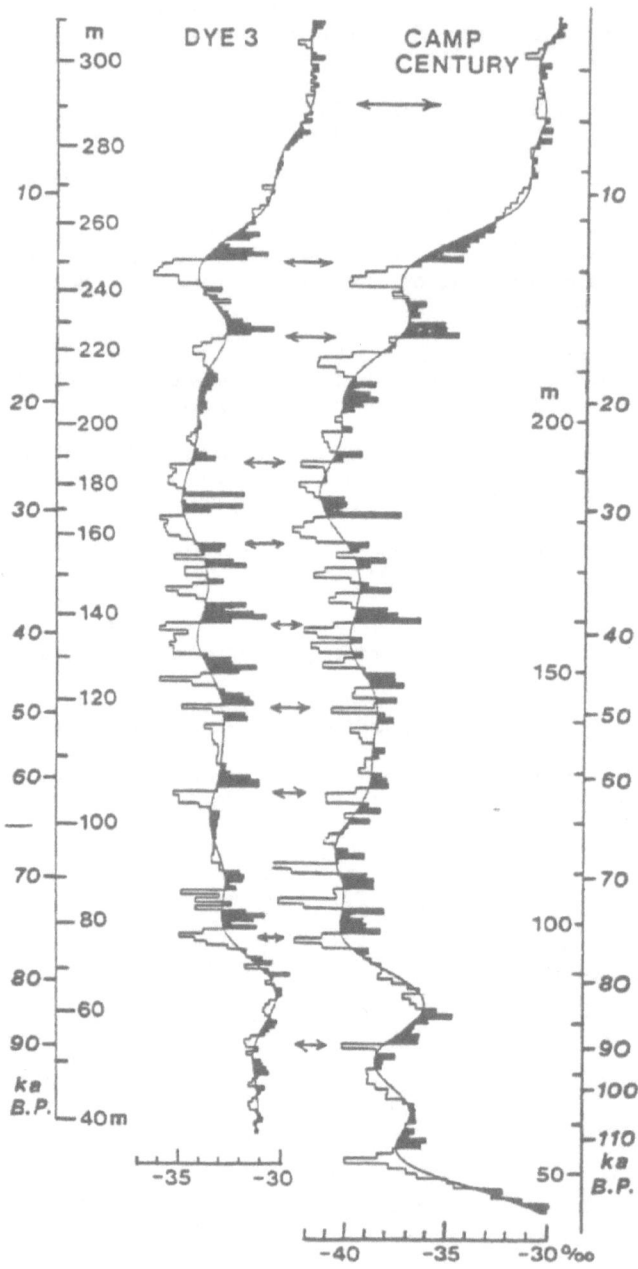

Fig. 1   Comparison of the deepest few hundred meters of two δ^{18}O profiles
through the ice sheet at Dye 3 in SE Greenland and at Camp Century
in NW Greenland.

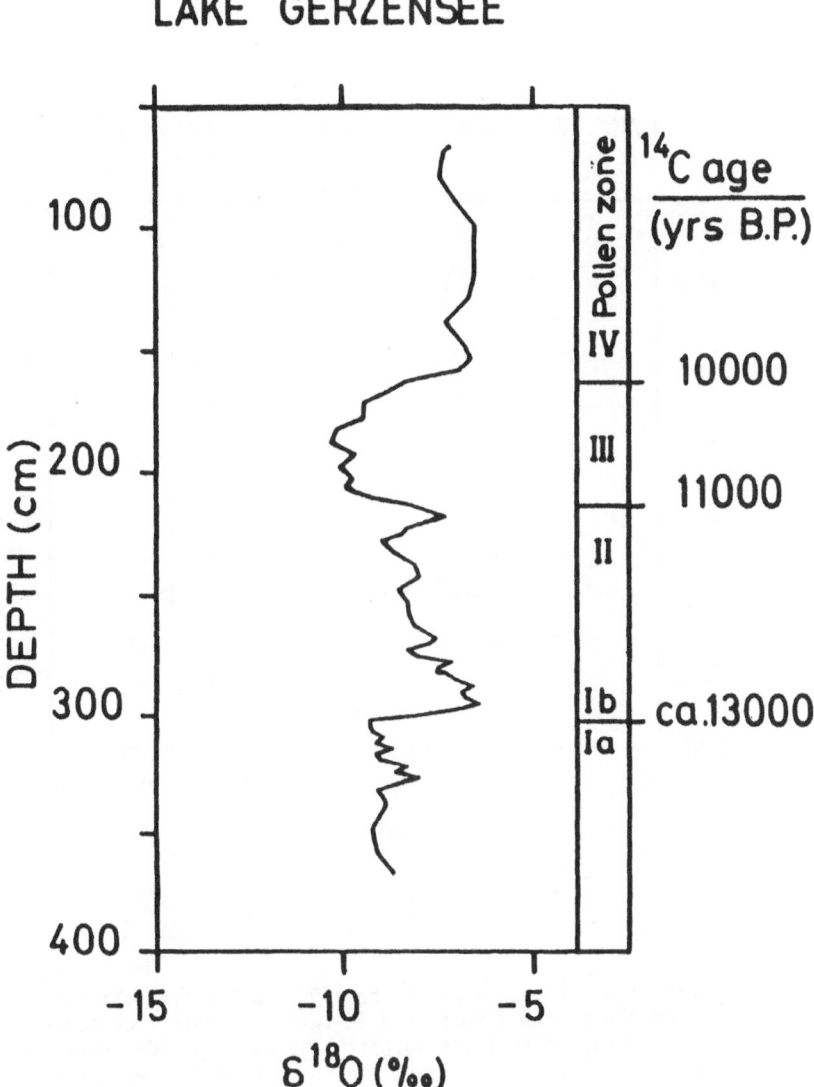

Fig. 2 $\delta^{18}O$ vs. depth in marl sediment of Lake Gerzensee, Switzerland, (from Eicher, 1980). The profile section covers the Late Glacial and Early Postglacial period, see approximate time scale at the right. The three pollen zone boundaries at ca. 13,000, 11,000 and 10,000 B.P. are marked by strong ^{18}O shifts.

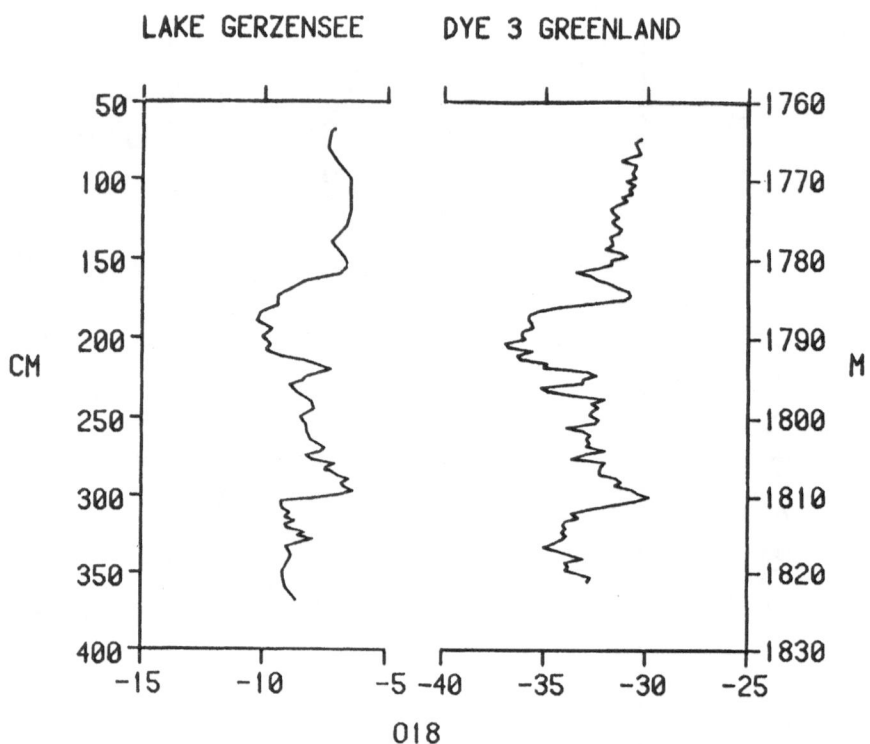

Fig. 3 Comparison of a section of the $\delta^{18}O$ profile from the Dye 3 ice core
(right) with the $\delta^{18}O$ record in Lake carbonate from Gerzensee (Left).
The strong similarities suggest that both records represent the same
sequence of climatic events and thus the same time period.

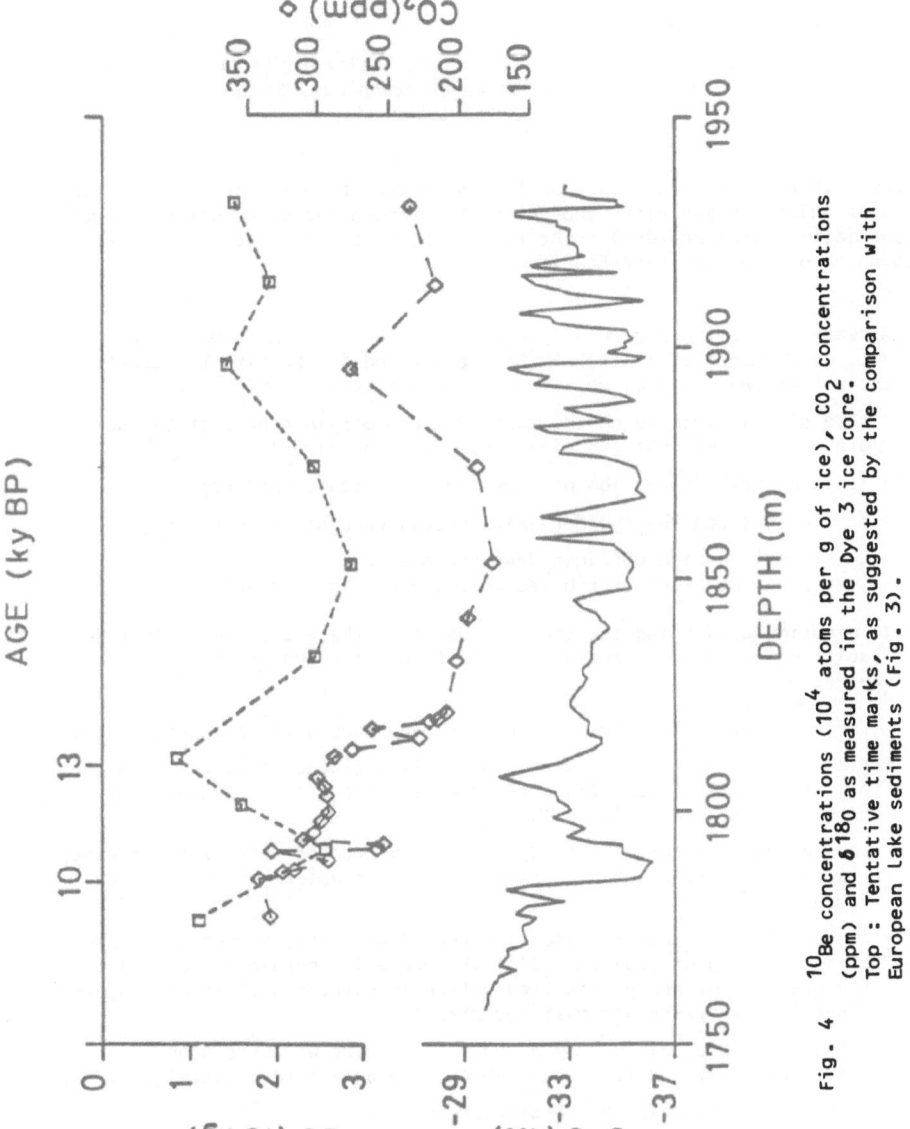

Fig. 4 ^{10}Be concentrations (10^4 atoms per g of ice), CO_2 concentrations
(ppm) and $\delta^{18}O$ as measured in the Dye 3 ice core.
Top : Tentative time marks, as suggested by the comparison with
European Lake sediments (Fig. 3).

POLLENANALYSES AND CHARACTERS OF CLIMATIC CHANGES AT
THE END OF THE EEMIAN AND AT THE BEGINNING OF THE
LATE WURM IN WESTERN EUROPE

J.L. de BEAULIEU, J. CLERC, M. COUTEAUX, A. PONS et M. REILLE

Laboratoire de Botanique historique et Palynologie (ERA 404
du CNRS), Faculté des Sciences et Techniques St-Jérôme,
13397 Marseille Cedex 13.

Theme A chosen for this symposium led us to gather some remarks about two
periods which are generally thought to have known sudden climatic changes
(especially in temperature) : the end of the last Interglacial and the
beginning of the last Late-Glacial.

I. A beautiful Eemian sequence, proposed for publication in Boreas by
J.L. de BEAULIEU et M. REILLE, has been found in the site Les Echets, an
ancient filled up lake, 15 km north-east of Lyon :

- this site is located on the south-western morainic part of the Dombes
  plateau (between the Ain river and the Saône river);

- it is a small depression probably dug by a tongue of ice;

- it lies between the two morainic systems recognized in the region :

  . internal würmian moraines down the plateau,
  . external moraines attributed to the penultimate glaciation.

Information concerning all the periods after the end of this penultimate
glaciation could be expected from the lacustrine and pet filling of this
depression.

A large diameter, 58 m deep boring has been made down to a sandy bottom :

- first pollen-bearing sediments show a Pleniglacial followed by a short
  climatic amelioration (Pinus and Jupiperus) and a reappearance of
  herbaceous steppe taxa;

- then a long complete forest cycle in evidenced in which some floristic
  elements (Taxus, Buxus) indicate that it undoubtedly belongs to the
  Eemian;

- above, after a glacial episod (increased Artemisia percentages) marked
  by a reworking of sediments from the preceding perdio accounting for
  the presence of mesophytic tree pollen in spectra with Artemisia, two
  other forest cycles are distinguished :

  . the first one, divided by a short clod episode, corresponds to
    St-Germain Ia and St-Germain Ib of la Grande Pile (WOILLARD, 1978),

  . the second corresponds to St-Germain II.

The very rare occurrence of Buxus, the absence of Hedera and Viscum and
unachieved successive dynamics lead us to assimilate these episodes to
the great interstadia of the Early Würm.

Then begins a long period of general unsteadiness well reflected through
A.P. fluctuations and during which three episodes may be considered as
interstadia : it corresponds to the Middle Würm.

There follows a long argilaceous sedimentation that yielded pollen spectra suggesting a very arid and cold steppe vegetation, and in which three organic layers made possible some C14 datings which confirmed that it belongs to the Late-Würm.

A great homogeneity at the end of the sequence enables to distinguish two zones :

- under 4 m, Pinus values oscillate around 30 %, with some aleatory peaks, those of Artemisia approximating 10 %;

- above 4 m, Pinus values range between 15 and 20 % (with three more or less aleatory peaks) and those of Artemisia are around 25 %.

At 1 m, there is a hiatus separating this Würmian from a boreal peat, but at the center of the present peat-bog, the beginning of the Bölling climatic amelioration is noticed on top of this Würmian, which therefore belongs to the Oldest Dryas.

The end of the Eemian should be examined first.

What caracterizes it is that the passage from a temperate forest with Abies and still abundant Quercus to a boreal forest with Pinus and Betula takes place in a very complex manner : first, the decline of Abies favours Picea and Pinus, the latter reaching very high values, then there occurs a sudden and strong decrease of Pinus, while Picea values remain unchanged and Abies has a short revival. Lastly takes place the last optimum of Pinus which predominates over the other conifers : it is the prelude to Betula expansion, followed by an herbaceous vegetation.

However, the abrupt aspect of the curves and rebedding phenomena is the overlying glacial layers may raise doubts as to the real existence of this complex history. Two verifications allow one to remove these doubts (BEAULIEU and REILLE, unpublished).

- a detailed study of the core revealed a perfectly homogeneous guttja (contrary to what was observed in overlying glacial layers);

- the number of spectra was increased as well as that of the samplings (32 instead of 14) :

  . the first maximum of Pinus is confirmed by six levels and by the general aspect of the curve,

  . during the transitory decline of Pinus, 12 levels enable to verify Picea stability and Abies revival.

Therefore, it is possible to question the rapidity of the climatic deterioration that took place at the end of the Eemian, such as it has been emphasized by G. WOILLARD (1979).

As a support to our statement, we found in litterature three examples of quite similar curves for Pinus (Fig. 1) :

- at the very site of la Grande Pile : in the diagram from boring X (and not boring III which has been analysed in the study just referred to), the first peak of Pinus rests on two facts and the phenomena accompanying Pinus pause are the same as at Les Echets (with slightly different percentages) : maintained values of Picea, reappearance of Abies; all this is quite normal, since La Grande Pile is only 200 km distant from Les Echets;

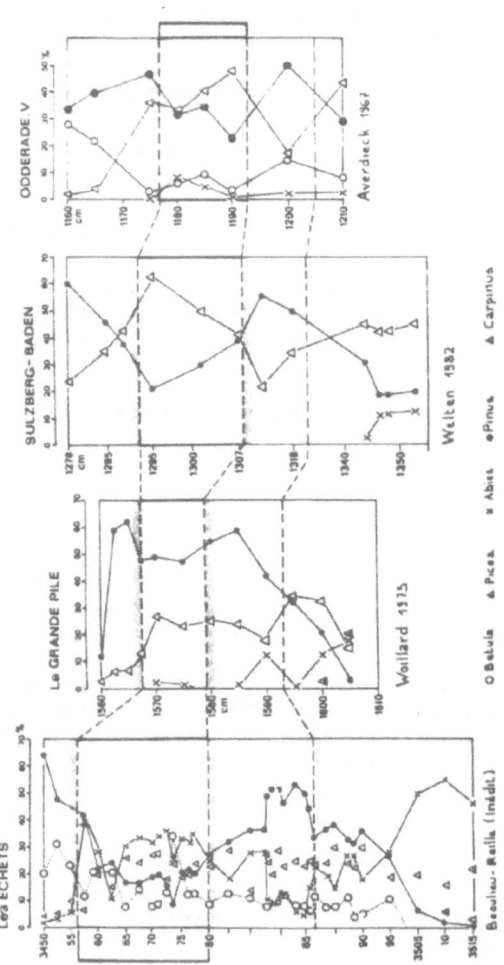

FIGURE 1

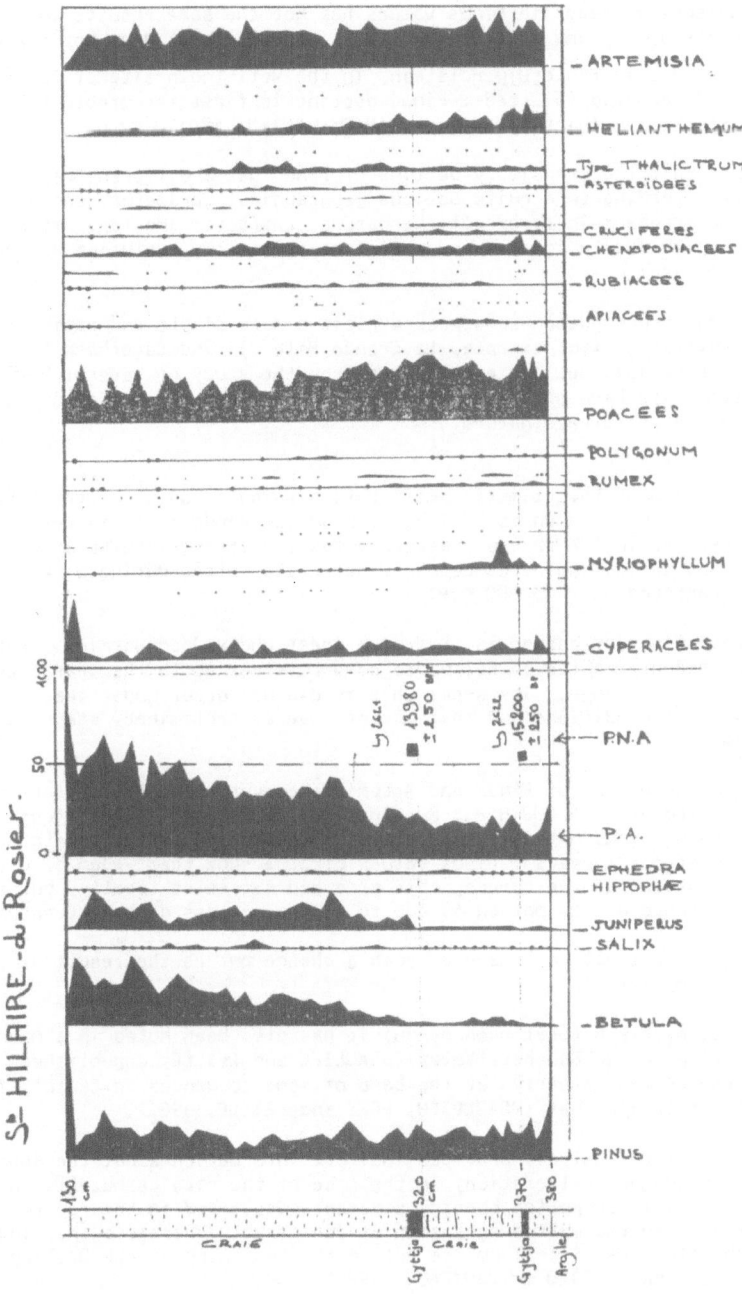

FIGURE 2

- in Zurich Oberland, at Sulzberg-baden, the same twofold motion of Pinus can be recognized but, of course, in this more continental site, the transitory decrease in Pinus values has not the same results (only Picea benefits by it, and still rather late, whereas Abies does not reappear);

- northwards, in Schleswig Holstein, in the well-lnown site of Odderade, a third scenario is acted : Pinus decline is first favourable to Picea and, later, to Abies (AVERDIECK, 1967; WELTEN, 1981).

The fact that the transitory decline of Pinus has a different regional history according to a quite obvious geographical parameter, seems to be a strong argument in favour of the natural character and long extent of this complex vegetational - and therefore climatic - evolution at the end of the Eemian.

Of course, other pollen sequences exhibit a more simple and rapid vegetational evolution (for example, La Grande Pile III and Samerberg in south-eastern Bavaria), but our experiences from the study of several borings in a same site have shown us that more complete sequences are both quite rare and nearly always natural.

II. The beginning of the climatic amelioration which is shown by the Late-Würm interstadium known as Bølling-Allerød (or Windermere) is well characterized in French sout-eastern sites : a strong impulse of pioneer shrubs and trees, Juniperus-Betula-Pinus, takes place during a short period centered on (13 300 B.P.

The prevailing impression is that of a great and sudden warming. But several recent results compel us to some reserves as to the characters of this climatic change. Its seems that it did not occur under the pleniglacial conditions and that one or several preliminary steps have existed.

- The intercrossing of Pinus and Artemisia curves noted at Les Echets (and dated at - 15 000 B.P.) must not be interpreted as a reflect of the decline of Pinus but as the result of a thickening of the vegetal cover due to steppe species : Pinus values of 30 % show that, even before the intercrossing of the curves, this tree did not exist locally, but that the presence of its pollen is due to a more or less distant transport.

It is difficult not to interpret such a change but as the result of a climatic amelioration.

It is not merely a local phenomenon; it has also been noted in a Mackereth boring from Lac du Bouchet (Velay)(BEAULIEU and REILLE, unpublished) and even, though less clearly, at the base of some sequences in Cantal, for example at La Taphanel (BEAULOIEU, PONS and REILLE, 1982).

Besides, pollen analysis provides indirect information about the importance of this presumed amelioration, in the case of the site La Muzelle in Oisans : the first pollen bearing sediments deposited in this depression are older than the marked expansion of Juniperus. Therefore they indicate that the place was free from ice before the beginning of the Bølling ... at an altitude of 2140 m (COUTEAUX, 1983).

To this early climatic amelioration should be ascribed the long pause in the modest increase of Juniperus frequencies, that regurlarly appears in southern Alps sites such as Pelleautier and Siguret[*] (BEAULIEU and REILLE, unpublished).

Quite recently, a plain site (St-Hilaire du Rosier in the Lower Isère valley) (CLERC, unpublished, Fig. 2) exhibited similar facts : above an argilaceous - pleniglacial -sediment, a lacustrine marly sediment indicates an Oldest Dryas type steppe vegetation; then there occurs a sedimentary change (gyttja bed) resulting from a great accumulation of algae which probably suddenly pulluated in the lake. At the same time appear aquatic plants (Myriophyllum) or shore plants (probably a part of Rumex and Umbelliferae) pointing to a thermic amelioration. Two levels with particularly abundant algae respectively yielded the dates of 15 200 $\pm$ 250 B.P. and 13 980 $\pm$ 250 B.P. The second level marks the beginning of a slow increase in Juniperus values; however, only slightly above do they reach frequencies which could indicate the beginning of the Bølling. If this increased growth corresponds to the event that took place at 13 300 B.P., the following climatic history can be outlined : first warming toward 15 200 B.P. (with a culmination at 14 700 B.P.), second warming at 14 000 B.P. and third greatest warming toward 13 300 B.P.

CONCLUSIONS

Two conclusions may be inferred from these facts :

- the climatic deterioration at the end of the Eemian was affected by an oscillation and therefore was not particulary short;

- the marked climatic amelioration at the beginning of the Late-Glacial interstadial took place only after one or several steps relatively moderate improvement extending over a period which may have lasted 1500 to 2000 years.

REFERENCES

AVERDIECK, F.R., 1967. - Die Vegetationsentwicklung des Eem-Interglazials und der Frühwürm-Interstadiale von Odderade/Schleswig-Holstein. Fundamenta (B), 2, pp. 101-125.

BEAULIEU, J.L. de, PONS, A. and REILLE, M., 1982. - Recherches pollen-analytiques sur l'histoire de la végétation de la bordure nord du Massif du Cantal (Massif Central, France). Pollen et Spores, 24 (2), pp. 251-300.

COUTEAUX, M., 1983. - La déglaciation du vallon de la Lavey (Vallée du Vénéon, Massif des Ecrins, Isère, France). Société hydrotechnique de France, Section de Glaciologie, Réunion de Grenoble, mars 1983.

WELTEN, M., 1981. - Gletscher und Vegetation im Lauf der letzten hunderttausend Jahre. Vorläufige Mitteilung. Jb. Schweiz. ntf. Ges. wiss. Teil 1978, pp. 5-18.

WOILLARD, G., 1978. - Grande Pile peat bog : a continuous pollen record for the last 140 000 years. Quaternary Research, 9, pp. 1-21.

WOILLARD, G., 1979. - Abrupt end of the last interglacial s.s. in north-east France. Nature, 281, pp. 558-562.

[*]Here, a series of six dates unfortunately had to be rejected because they looked much exaggerated.

LES CRISES CLIMATIQUES DE COURTE DUREE (QUELQUES ANNEES A QUELQUES SIECLES)
ET LEUR ENREGISTREMENT DANS LA SEDIMENTATION CONTINENTALE

par Pierre ROGNON, Professeur à l'Université P. et M. Curie, Paris.

Les progrès récents des Sciences du climat ont amené, preque simultanément la
découverte d'accidents climatiques (abrupt events) qui surviennent de façon
aléatoire sur les courbes et se distinguent à la fois de la variabilité
"normale" du climat et des ondulations à long terme :

- en climatologie, ces anomalies ont été mises en évidence après prequ'un
  siècle d'observations qui avaient permis de définir la variabilité de chaque climat et la tendance globale vers un réchauffement jusqu'en 1940,
  puis un léger refroidissement ensuite. Sur ces courbes, la série d'hivers
  rigoureux de 1939-43 en Europe occidentale ou les sécheresses de 1968-73 au
  Sahel ou de 1979→1982 en Astralie orientale par exemple apparaissent comme
  des pics accentués, de courte durée et de faible récurrence (1 à 3-4 fois
  par siècle). Pour étudier ces anomalies, la durée des observations météorologiques (un siècle en moyenne) ne suffit pas : il faut de plus en plus
  recourir à la dendroclimatologie.

- en palaéoclimatologie, des crises de même nature apparaissent sur les
  courbes de reconstitution du climat, grâce à la précision atteinte par les
  techniques palaéoclimatiques. Ces crises ont les mêmes caractéristiques
  que les anomalies mais en diffèrent par leur durée qui varie de quelques
  siècles à près d'un millénaire sur des durées d'observation de milliers ou
  dizaines de milliers d'années. Ces crises ont pu être mises en évidence
  quand des études précises ont commencé sur les sédiments continentaux qui
  ont, sur les sédiments océaniques, un double avantage : une plus grande
  sensibilité aux moindres variations climatiques et une plus grande vitesse
  de sédimentation, qui permet une étude plus fine de ces variations.

Anomalies et crises climatiques présentent de curieuses convergences dans
leur "comportement" :

- à la différence des variations à long terme qui ont souvent une extension
  à l'échelle du globe ou d'un hémisphère, les anomalies et les crises sont
  circonscrites à une région, une portion de continent (Australie orientale,
  Europe occidentale), parfois une zone climatique (Sahel)

- elles ont une distribution très aléatoire dans le temps. Elles sont parfois inexistantes sur d'assez longues périodes de stabilité, mais peuvent
  aussi se grouper, entraînant une extrême instabilité très néfaste pour
  l'équilibre des milieux naturels et de la biomasse.

Cette constatation a une portée considérable à la fois pour l'avenir de nos
sociétés (rupture des équilibres économiques lors des anomalies thermiques
ou pluviométriques) et pour l'explication des modifications des milieux naturels dans le Passé. D'où l'idée de mener une étude conjointe des crises et
des anomalies sur les séries les plus longues possibles. Dans cet exposé,
nous allons chercher à répondre à la question : à quel type d'archives géologiques peut-on s'adresser ? Deux exemples sont présentés ici à partir des
dépôts lacustres et des tourbières, qui ont l'avantage de fournir des séquences continues, à fort taux de sédimentation (donc une grande précision
dans la chronologie relative) et qui reflètent fidèlement l'évolution du
climat des continents.

I. RECHERCHES INTERDISCIPLINAIRES SUR LES DEPOTS LACUSTRES DE L'HOLOCENE INFERIEUR AU SAHEL

Les régions sahéliennes sont caractérisées par une longue saison sèche et la concentration des pluies sur 2 à 3 mois. Au cours des derniers millénaires, ce rythme annuel a subi des variations dont témoignent les fluctuations de la limite nord du peuplement sédentaire. Mais faute de documents historiques nombreux (S. Nicholson, 1980) et d'études dendroclimatologiques (très difficiles en milieu tropical), les anomalies pluviométriques ou les crises sont encore très peu connues au-delà de la période couverte par les mesures météorologiques.

Par contre, à l'échelle des 10-12 derniers millénaires, on connaît bien la tendance climatique pour toute la zone sahélienne. Entre 10 ET 20° N, l'hyperaridité du climat qui régnait depuis 18-20 000 ans BP fait place à partir du 10 000 BP à un climat très humide qui favorise l'installation de lacs dans toutes les dépressions fermées et les creux interdunaires. Ces lacs souvent profonds avaient de faibles variations saisonniières, ce qui indique une bonne répartition des pluies sur la quasi-totalité de l'année. Ce pluvial a duré jusque vers 5000, 4500 BP, lorsque, de façon tout aussi rapide, la zone des pluies s'est retirée vers le Sud, entraînant l'installation des conditions de semi-aridité actuelles. Or sur cette courbe d'évoluation d'ensemble, on constate au moins deux périodes de crises :

1° Une crise d'extrême pluviosité vers la fin de la phase hyperaride (12-10 000 ans BP). Cette crise est encore très mal connue en dehors du bassin du Nil (fig. 1). Là, des crises d'une ampleur exceptionnelle (22 mètres au-dessus du lit acutel à Kom Ombo; 30 mètres à la frontière soudano-égyptienne) se sont produites peu après 12 000 BP (K.W. Butzer et C.L. Hansen, 1968). Sur une durée de quelques siècles, elles ont laissé des limons d'inondation (Darau Member), épais parfois de 3 mètres, très au-delà de la vallée actuelle du fleuve et jamais depuis, le Nil n'a connu de telles crues. Cette période de crues exceptionnelles est confirmée par J. de Heinzelin (1968) ou F. Wendorf et R. Schild (1976). Ces apports d'eau douce considérables ont entraîné l'installation d'un fort gradient de salinité en Méditerranée orientale et la sédimentation d'un sapropel entre 11 760 et 10 440 BP. Or l'étude des $\delta^{18}O$ des carbonates sur Globigerinoides ruber indique un pic négatif de plus de 1,5 % juste avant $11\ 760 \pm 142$ ans BP (M. Rossignol - Strick et al, 1982). Cette crise pluviométrique n'est, pour l'instant, pas connue en dehors de la vallée du Nil. Elle est à mettre en parallèle avec les crises thermiques qui caractérisent la fin du Tardiglaciaire en Europe occidentale.

2° Une crise d'aridité a, par contre, été mise en évidence sur toute l'étendue de la zone sahélienne depuis l'Atlantique jusqu'à la Mer Rouge. Elle interrompt pour quelques siègles ou un millénaire peut être, le Pluvial de l'Holocène inférieur à moyen (fig. 2). Elle se marque par une discontinuité dans la succession des dépôts lacustres de cette période, aussi bien dans la cuvette tchadienne (M. Servant, 1973), qu'en Afar ou même sur les hauts plateaux éthiopiens (F. Gasse et al, 1980). On constate, en Afrique de l'Ouest, la disparition de nombreux lacs, avec une reprise des dépôts dunaires en Mauritanie orientale (Tichitt)(G. Hugot, 1977), ou de l'ablation éolienne au N.W. de l'Adrar des Iforas (Erg In Sakane) au Nord du Mali (C. Hillaire Marcel et al, 1983). Au Sénégal, les dunes fixées "ogoliennes" datant de la période

antérieure à 10 000 ans sont partiellement remaniées lors de cette
crise (Leprun, 1971) et le fleuve Sénégal dépose le "deuxième remblai"
d'alluvions le long de sa vallée à la suite de la diminution de son
débit (P. Michel, 1973).

Les premières datations ^{14}C situaient cette crise dans une fourchette de
temps autour de 8 400 - 6 800 BP au Sénégal, entre 8 400 et 7 500 BP dans
l'Afar ou vers 7 500 BP au Tchad et l'on pensait que cet événement avait
été synchrone à l'échelle de la zone sahélienne. En effet cette crise,
dans tous les secteurs, marque une modification fondamentale des mécanis-
mes de la pluie.

- Avant cette crise, l'étude des pollens, des diatomées, des sédiments flu-
  viatiles ou lacustres indique un étalement indiscutable des pluies sur
  l'année. Les diatomées, au moins au Tchad, indiquent des températures
  inférieures à l'Actuel. L'économie préhistorique est alors fondée sur
  l'exploitation des mammifères aquatiques et des poissons depuis la Mauritanie jusqu'au Nil. Le "hiatus" dans l'évolution des sociétés humaines
  est placé après 7 500 et avant 6 000 BP (A.B. Smith, 1980).

- Après cette crise, le régime des pluies est typiquement celui de la
  mousson, mais avec une saison pluvieuse nettement plus longue qu'aujourd'
  hui puisqu'on observe à nouveau des lacs ou des marécages et en dehors
  des milieux hydromorphes, formation de sols. Sur ces sols, de nouvelles
  populations pratiquent désormais l'élevage des bovins, des moutons et
  des chèvres et l'agriculture sédentaire. La fin de ce Pluvial, vers
  4 500 BP correspond en fait à une simple réduction de la durée de la
  saison pluvieuse, responsable de l'assèchement généralisée du climat.

Toutes les données tirées des paléoenvironnements ou des genres de vie
préhistoriques indiquent donc très nettement une discontinuité ou un
hiatus en relation avec une crise climatique que l'on plaçait vers 7 500 -
7 000 BP. Cette crise correspond à un changement important dans la répartition saisonnière des pluies avant et après la crise. Or des travaux
récents semblent démontrer que cette crise n'est pas synchrone sur toute
la zone sahélienne :

- au NW des Iforas, une série de datations ^{14}C localisent cette crise
  entre 6 400 et 5 400 BP, c'est-à-dire vers la fin du Pluvial de l'Holocène inférieur et moyen, au cours duquel le niveau des nappes semble
  avoir toujours été élevé (G. Hillaire Marcel et al, 1983). Cette crise
  "tardive" affecte curieusement la région la plus au Nord (20°50 N) donc
  la plus proche du désert alors que cette crise serait plus précoce au
  Sud : Sénégal (15-17° N), Tichitt (18° N) ou Tchad (10-10° N) ?

- au Rajastan, à l'extrémité asiatique de la même zone sahélienne, une
  crise d'aridité du même type se place entre une longue période holocène
  caractérisée par des pluies à la fois de saison chaude et de saison
  fraîche (d'après les études polliniques) et la période pluviale plus
  récente où ne subsistent que des pluies de mousson. Cette crise se
  situe entre 5 000 et 3 500 BP environ et correspond à un épisode d'assèchement des lacs et à la disparition de la civilisation de l'Indus
  (Bryson et Swain, 1981).

Ainsi, dans l'état actuel des connaissances, on observe à travers toute
la zone sahélienne un événement exceptionnel que l'on peut appeler une
crise climatique. Un gros effort de datation devrait être fait pour pré-

ciser sa durée et surtout sa date exacte d'un bout à l'autre de la zone. En effet, la signification d'un tel phénomène sera très différente s'il est synchrone ou s'il est décalé selon la latitude ou la longitude.

II. ETUDES ISOTOPIQUES SUR LES TOURBIERES DES REGIONS TEMPEREES

Les tourbières en régions tempérées sont utilisées depuis très longtemps comme archives palaéoclimatiques pour les analyses polliniques. Elles conservent parfaitement les pollens dans ces milieux hydromorphes et ces pollens ont généralement une origine locale. Grâce à l'analyse des pollens de tourbières, on a pu mener des études précises en reconstituant, par exemple, certains interstades de la dernière glaciation. Mais peut-on utiliser les pollens des tourbières pour reconstituer des épisodes aussi brefs que les crises ? Théoriquement, cela est possible sur les tour-bières à grande vitesse de sédimentation de l'ordre d'un mètre/millénaire. Pourtant Jacobson et Bradshaw (1981) ont comparé la variabilité des dia-grammes polliniques d'une tourbière et d'un lac voisins de Finlande. Les pics obtenus sont effectivement très marqués sur la tourbière et très at-ténués ou absents sur le diagramme du lac. A priori donc, la tourbière paraît un meilleur marqueur des crises brèves alors que celles-ci sont masquées dans les sédiments lacustres qui rassemblent les apports fluvia-tiles d'un vaste bassin versant. Pourtant Jacobson et Bradshaw soulignent que ces pics, dans les tourbières, peuvent parfois indiquer, soit une sur-représentation de certaines expèces très proches (voire implantées sur la tourbière) à certaines époques, soit des variations rapides dans les vitesses de sédimentation, ce qui est fréquent dans les tourbières en fonction de leur régime hydrique. Si la tourbe est plus ou moins tassée ou aérée, le pourcentage de pollens par unité de volume s'en trouve modi-fié. Les pics très brefs observés sur des diagrammes devraient donc être précisés par une méthode d'analyse différente, afin d'être confirmés.

Or les tourbières n'ont pas seulement un rôle de réceptacle des pollens. Elles sont constituées par une matière organique qui a pu enregistrer un certain signal isotopique au moment de sa formation et le conserver en-suite. Les études menées depuis 1975 par A. Ferhi au sein de notre équipe ont, en effet, montré qu'il exite une relation étroite entre les teneurs en ^{18}O de la matière organique et l'environnement climatique. Ceci a été vérifié sur les plantes actuelles (A. Ferhi et R. Letolle, 1979), sur des végétaux cultivés en milieu contrôlé (A. Ferhi et al, 1976; A. Ferhi et R. Letolle, 1977) sur les litières des sols de type podzol (M. Balabane, 1978) et enfin sur des tourbes actuelles (P. Gégout, 1982). Une relation existe avec la température par l'intermédiaire de la composition isotopi-que de l'eau du sol, de celle de la vapeur d'eau atmosphérique et enfin de l'humidité relative de l'air. Cette relation varie donc en fonction de l'altitude et de la latitude qui modifient la composition isotopique de l'eau. Mais à priori, il est possible d'étudier, sur une même carotte, les variations thermiques des dernières dizaines de millénaires.

Cette méthode d'analyse a donc été appliquée à l'étude de quelques tour-bières à Longeroux sur le plateau de Millevaches dans le Massif Central et à Venise (fig. 3). Des tests ont été effectués pour s'assurer que les pics obtenus sur les courbes isotopiques n'étaient pas dus à un manque d'homo-généité de la tourbe (test de reproductibilité) ou à des variations dans l'état de conservation de la tourbe (rapport C/N, $\delta^{13}C$ ou $\delta^{15}N$). Ces pics ont ensuite été corrélés avec ceux qui étaient fournis par les ana-

lyses polliniques. Les résultats obtenus, malgré une maille d'échantil-
lonnage entre très grossière, se révèlent concordants. Ils ont été dis-
cutés dans une publication récente (A. Ferhi et al, 1982).

Certains avantages de cette méthode isotopique sont à retenir :

- Les teneurs en ^{18}O augmentent chaque fois que les analyses polliniques
  indiquent une "amélioration" des conditions climatiques et inversement
  à la fois pour la fin du Pleistocène supérieur (Venise) et pour
  l'Holocène (Longéroux). A Venise, les variations obtenues sont, par
  exemple, de 7 %$_o$ à l'intérieur de la période 14 000 à 40 000 BP, ce
  qui est à rapprocher des différences obtenues dans l'Actuel entre les
  végétaux de régions méditerranéennes et scandinaves (de l'ordre de 9 %$_o$).

- Les mesures sont effectuées sur la matière organique elle-même dont
  elles traduisent les réactions physico-chimiques ou biologiques. Elles
  sont donc indépendantes des modifications du couvert végétal par les
  défrichements à l'époque néolithique. A Longéroux, on constate que
  la courbe paraît rester fiable même après le début du Subboréal.

- La quantité très faible de matière nécessaire (quelques grammes ou
  milligrammes) permettrait de réserver la maille des mesures jusqu'à
  l'échelle du siècle, qui est celle des crises climatiques.

Mais cette méthode isotopique se heurte encore à certains obstacles :

- La relation avec les paramètres thermiques doit être précisée par une
  analyse approfondie des causes des fractionnements isotopiques aux différentes étapes de la formation et de l'évolution de la matière organique. Cette condition est nécessaire pouré tablir des fonctions de
  transfert fiables.

- Le choix de la tourbe doit être rigoureux. Ainsi à Venise, la courbe
  isotopique n'est utilisable qu'entre 40 000 et 14 000 lorsqu'une
  tourbe à Cypéracées très homogène s'accumule sur place dans une dépres-
  sion en voie de lente subsidence. Mais après 14 000 BP, la tourbière
  reçoit de la matière organique extérieure, apportée par les eaux de
  fonte des glaciers alpins, puis après 7 000 BP, de la matière organique
  marine en rapport avec la transgression flandrienne.

- Au sein même de la tourbière, il est possible que le mélange de plantes
  aquatiques différentes ou, surtout, de plantes aquatiques et de plantes
  vasculaires lors de brèves périodes d'assèchement, risque de modifier,
  localement, la composition isotopique de la tourbe au moment de sa
  formation.

De nombreuses mises au point sont donc encore nécessaires. Mais cette
méthode isotopique est appelée, semeble-t-il, à jouer un rôle comparable,
par rapport aux analyses polliniques, à celui que remplit l'étude isotopique ^{18}O sur les carbonates des Foraminifères, parallèlement aux déter-
miniations paléontologiques des associations de Foraminifères dans le
milieu océanique. Elle pourrait être une méthode d'approche indépendante
et apporter des informations complémentaires qui restent encore à définir.

CONCLUSION

La mise en évidence de crises et anomalies ouvre une voie nouvelle dans les recherches sur l'évolution du climat. Dans la phase actuelle, il s'agit de mettre au point de nouvelles méthodes plus précises pour recenser ces "abrupt events" à toutes les échelles de temps. Dans une étape ultérieure, on peut espérer trouver une logique dans la répartition temporelle ou spatiale de ces événements, même si celle-ci nous apparaît actuellement aléatoire. Le but à long terme est d'arriver à prévoir les anomalies climatiques qui sont à l'origine des catastrophes naturelles d'origine climatique (P. Rognon, 1982). Ces catastrophes sont en effet celles qui affectent les plus vastes zones sinistrées (à l'échelle de portions de continents ou de zones comme le Sahel) et qui causent, sinon le plus grand nombre de victimes, mais les plus grandes pertes économiques à l'échelle mondiale. Leur prévision, qui passe par une meilleure connaissance des anomalies et des crises, devrait être une des priorités de tout programme de recherches sur l'évolution du climat.

REFERENCES

Balabane M., (1978). - Reconnaissance des variations de la composition isotopique (^{18}O ^{16}O) de la matière organique accumulée en surface des podzols dans différents environnements bioclimatiques. Thèse doc. 3e cycle, Univ. P. et M. Curie, Paris, 54 p.

Bryson R.A. et Swain A.M., (1981). - Holocene variations of Monsoon rainfall in Rajasthan. Quat. Res., 16, pp. 135-145.

Butzer K.W. et Hansen C.L., (1968). - Desert and River in Nubia, geomorphology and prehistoric environments at the Aswan Reservoir. Univ. of Wisconsin Press, 584 p.

Ferhi A., Gegout P. et Rogonon P., (1982). - Premiers résultats palaéoclimatiques fournis par l'analyse de l'oxygène 18 dans la matière organique des tourbes. Bull. Ass. Fr. Et. Quat., 1, pp. 47-51.

Ferhi A. et Letolle R., (1977). - Transpiration and evaporation as the principal factors in oxygen isotope variations of organic matter in land plants. Physiol. veg., 15(2), pp. 363-370.

Ferhi A. et Letolle R., (1979). - Relation entre le milieu climatique et les teneurs en oxygène-18 de la cellulose des plantes terrestres. Physiol. veg. 17(1), pp. 107-117.

Ferhi A., Lang A. and Lerman J.C., (1976). - Stable isotops of oxygen in plants : a possible palaeohygrometer. Hydrology and water ressources in Arizona and the Southwest. Proc. of the 1976 met. of the Arizona section. American water ressources ASSN and the hydrology section. Arizona Academy of Science, Tucson. 6, pp. 191-198.

Gegout P., (1982). - L'oxygène-18 dans la matière organique des tourbes : relation avec les palaéoclimats. Thèse doc. 3e cycle, Univ. P. et M. Curie, Paris, 117 p.

Gasse F., Rognon P. et Street A., (1980). - Quaternary History of the Afar and Ethiopian Rift Lakes in "The Sahara and the Nile" MAJ Williams et H. Faure édit., Belkema, pp. 361-400.

Heinzelin J. de (1968). - Geological history of the Nile Valley in the Prehistory of Nubia, F.H. Wendorf edit., Southern Methodist Univ. Press, Dallas, 1, pp. 19-55.

Hillaire-Marcel C., Riser J., Rognon P., Rosso J.C. et Soulié-Marche I. (1983). - ^{14}C chronology of Holocen hydroclimatic changes in North Estern Mali. Quat. Res. (à paraître).

Hugot G., (1977). - Un secteur du Quaternaire lacustre mauritanien : Tichitt Inst. Maurit. Rech. Sc., 190 p.

Jacobson G.L. et Bradshaw R.H. (1981). - The selection of sites for palaeovegetal studies. Quat. Res., 16, pp. 80-96.

Leprun J.C. (1971). - Nouvelles observations sur les formations dunaires fixées du Ferlo nord occidental. Bull. Ass. Sén. Et. Quat., 31-32, pp. 69-78.

Michel P., (1973). - Les bassins des fleuves Sénégal et Gambie, étude géomorphologique. Mém. ORSTOM Paris, 63, 3 t, 752 p.

NIcholson S., (1980). - Saharan climates in historic times in the Sahara and the Nile, MAJ Williams et H. Faure edit., Balkema, Ritt., pp. 173-200.

Rossignol-Strick M., Nesteroff W., Olive P. et Vergnaud-Grazzini C., (1982). After the deluge : Mediterranean stagnation and sapropel formation Nature, 295, pp. 105-110.

Rognon P., (1982). - Les catastrophes naturelles d'origine climatique. Le Monde, Paris, 27 oct., pp. 13-14.

Servant M., (1973). - Séquences continentales et variations climatiques. Evolution du bassin du Tchad au Cénozoïque supérieur. Thèse doc. Et., ORSTOM, Paris, 368 p.

Smith A.B., (1980). - The Neolithic tradition in the Sahara in the Sahaea and the Nile, MAJ. Williams et H. Faure edit., Balkema, Rott., pp. 451-465.

Wendorf F. et Scmild R., (1976). - Prehistory of the Nile Valley. Acad. Press, N.Y., 404 p.

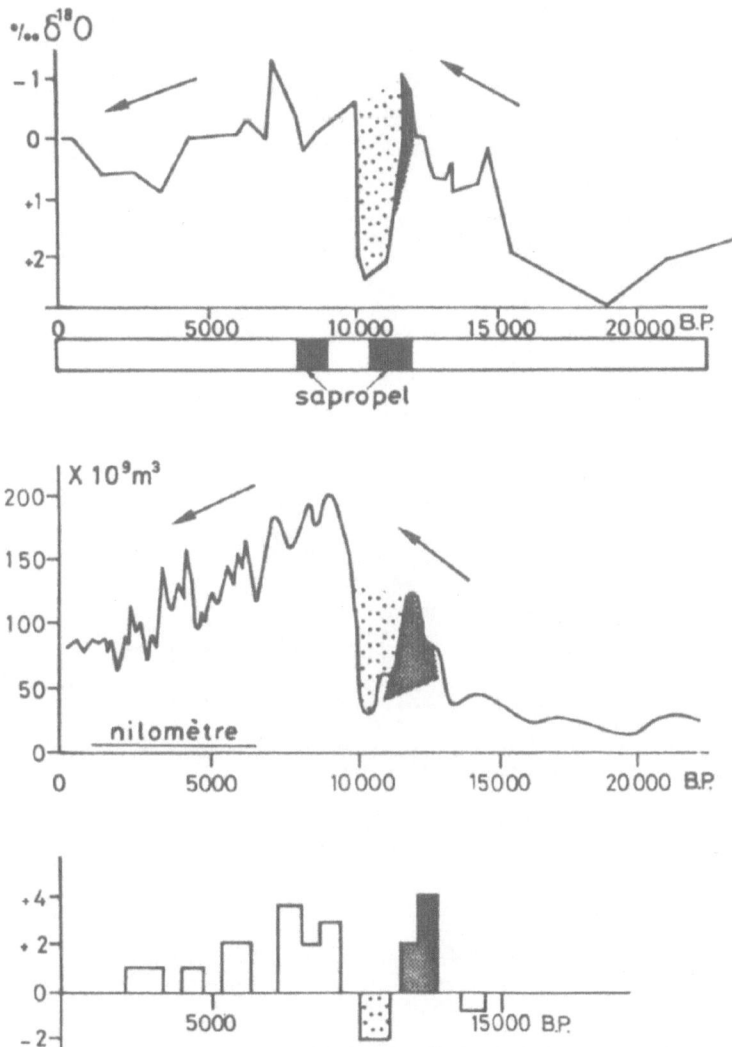

Fig. 1 : Crise d'extrême pluviosité dans le bassin du Nil : A. Dans le Soudan Central où les débits estimés s'élèvent à près de quatre fois le débit actuel pour la période vers 12 000 BP (M.A.J. WILLIAMS et D.A. ADAMSON, 1980, p 301); B. En Nubie où un pic bien individualisé apparait vers 12 000 BP (R.W. FAIRBRIDGE, 1976); C. En Méditerranée au large de l'Egypte, sur la carotte KS 52, un pic négatif de plus de 1,5°/₀₀ pour les δ ^{18}O sur Globigérinoïdes ruber apparait entre 12 600 et 11 700 BP et coïncide avec un niveau de sapropel dans la sédimentation marien.

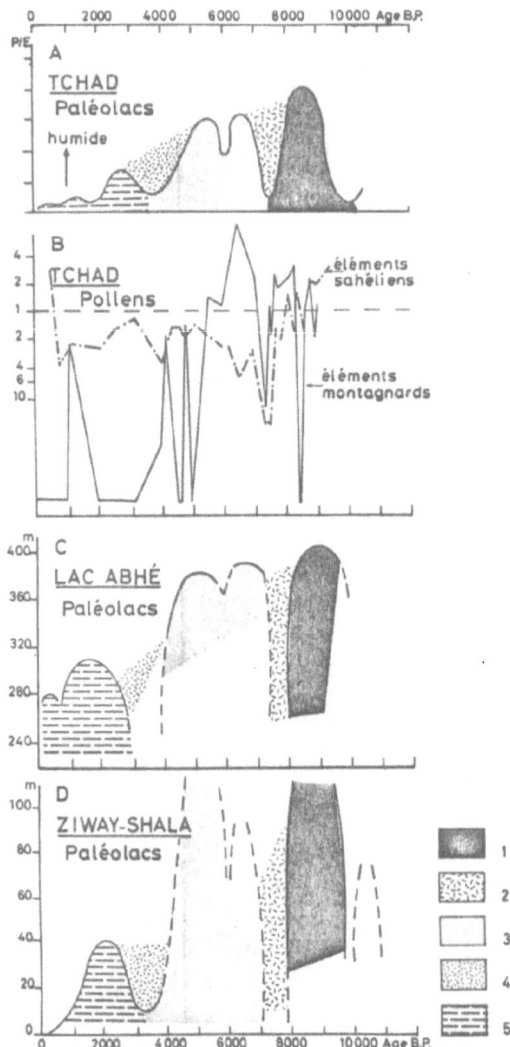

Fig. 2 :   Crise d'aridité en domaine sahélien : A. Rapport P/E d'après les
variations de niveau des paléolacs de la cuvette tchadienne
(M. SERVANT, 1973); B. Evolution des courbes polliniques des éléments montagnards et sahéliens dans la coupe de Tjéri au Lac Tchad
(J. MALEY, 1980); C. Variations des niveaux lacustres du lac Abhé
(République de Djibouti et Ethiopie) d'après les altitudes abso-
lues des paléorivages (F. GASSE, 1975); D. Variation des niveaux
lacustres du système Ziway-Shala (Ethiopie) d'après l'altitude relative des paléorivages par rapport au lac actuel (A. STREET, 1979).
Les trames indiquent : 1- pluies bien réparties sur l'année; 2-
crise d'aridité vers 8 000-7 000 BP; 3- pluies estivales abondantes; 4- aridité vers 4 500-4 000 BP; 5- pluies de mousson en ré-
gression.

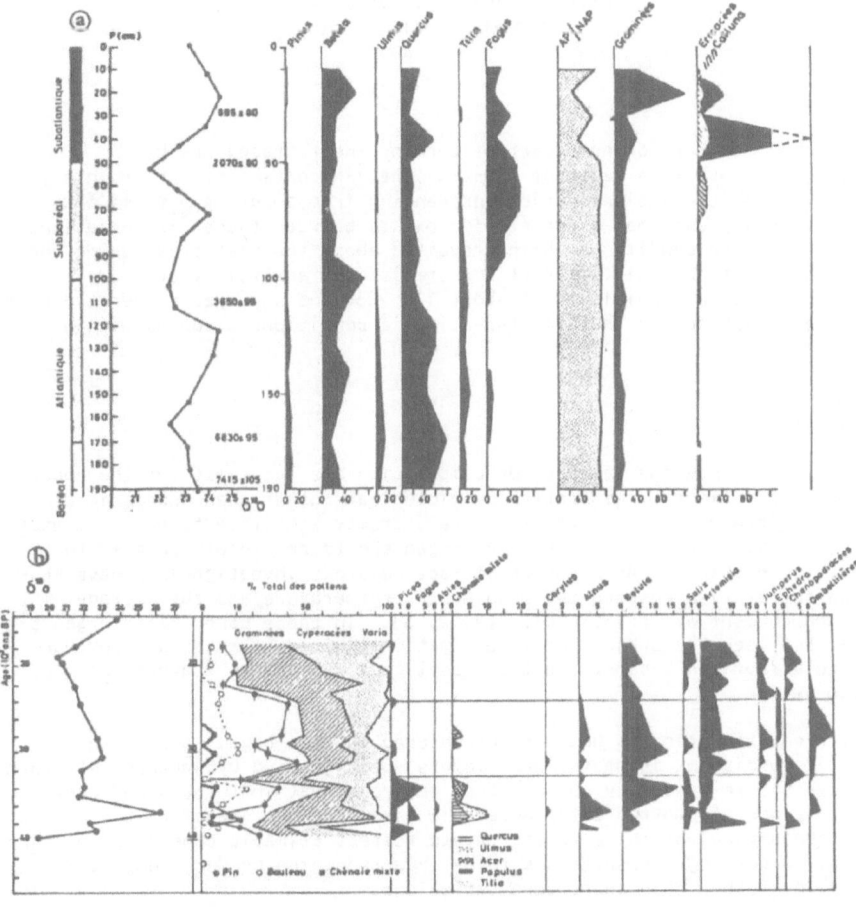

Fig. 3 : Comparaison des diagrammes polliniques simplifiés et des courbes
de variation de $\delta^{18}O$ dans les tourbières : A. à Longeroux (plateau
de Millevaches, Massif Central); B. dans une tourbière de Venise
(A. FERHI, P. GEGOUT et P. ROGNON, 1982).

Do ^{15}N Variations in Peat Bogs allow Statements
of Climatic Changes in the Past ?

G. H. Schleser
ICH-Abt. Biophysikalische Chemie
Kernforschungsanlage Jülich GmbH
D-5170 Jülich 1

Abstract

A peat bog from the northern part of Germany (near Stade) has been
investigated down to a depth of 440 cm. The ^{15}N content of the organic
material exhibits distinct variations ranging from about -4.5 to +4.5 %$_o$.
The results suggest that a correlation exists between these ^{15}N variations
and the climatic conditions which prevailed above the peat area during the
formation of its organic material. A statistical analysis yields cyclic
variations with periodicities of about 170, 280 and 1000 years over the past
4500 years, indicating that similar climatic conditions occurend within
these intervals.

Introduction

Nitrogen abundance and its isotope composition are, at least for the upper
lithosphere, primarily controlled by biological rather than inorganic processes. Since biological processes are extremly sensitive to environmental
changes, soil-organic matter and -nitrogen should be closely related to
climatic conditions. As a matter of fact numerous investigations have shown
a good correlation between the mean annual temperature and the average
nitrogen content of various soils [1, 2, 3]. In these cases total organic
matter has been the subject of investigations insted of material from one
particular horizon because the total soil body represents a dynamically
active system.

A different situation is however encountered in peat bogs. Due to the
specific conditions in peats, peat layers, being buried by successively formed
new layers, are gradually cut off from environmental influences and cease to
decompose organic matter after a certain time period. Therefore the degree
of decomposition for these layers should reflect climatic conditions of that
period. Accordingly results are to be expected which produce long term
tendencies rather than short term variations.

Besides it is known that during decay when complex organic matter is broken
down and smaller organic molecules are extracted from it, an ^{15}N enrichment
of the remaining complex organic matter is encountered. Therefore the
degree of humidification and the variation of ^{15}N could represent quantities
which allow statements of past climatic changes.

The present paper reports on first results with respect to ^{15}N.

Material and Methods

A peat bog near Stade (9°28'E; 53°44'N) lower Saxonia, had been chosen. Peat
samples were taken down to a depth of 440 cm. The various soil horizons have
been restricted to a width of 5 cm each. This procedure minimizes uncertainties which may exist due to variations in a horizon itself.

The cut samples were dried for about 2 days at 60° C and afterwards
thoroughly homogenized. Application of Kjeldahl's method yielded the nitrogen content of each sample. The Kjeldahl procedure was also applied to
prepare samples for ^{15}N analysis. Oxidation of Ammonia by Sodium-hypobromite
was used to introduce nitrogen into the mass spectrometer. The mass spectrometer analysis' have been done at Los Alamos National Laboratory.

The concentrations of ^{15}N are given as permil (o/oo) differences from
atmospheric molecular N_2 and are expressed in the delta notation :

$$\delta^{15}N = \left[\frac{(^{15}N/^{14}N)_{sample}}{(^{15}N/^{14}N)_{air}} - 1 \right] \times 1000$$

Results and Discussion

The nitrogen content and corresponding $\delta^{15}N$ values are shown Fig. 1, 2 as
function of age at which the corresponding layers have been formed. It is
noteworthy to indicate that both curves exhibit a long time trend. According
to this trend the values tend to decrease at the beginning somewhat while an
increase is exhibited with older soil layers i.e. deeper soil horizons (see
Fig. 3). If this trend is eliminated a cyclic pattern remains for the $\delta^{15}N$
curve which is likewise evident from Fig. 2.

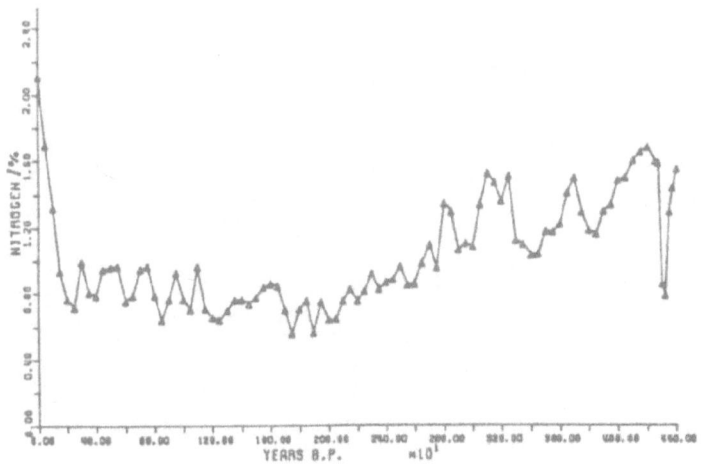

Fig. 1 : Organic nitrogen accumulation in a north German peat bog as a
function of its time of formation

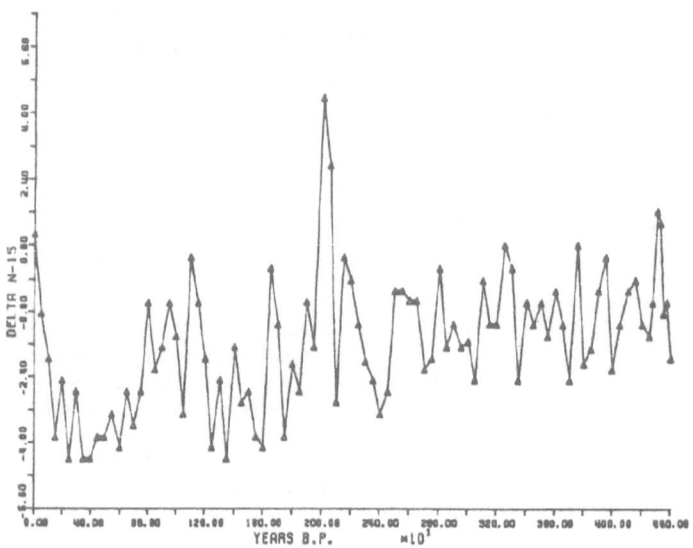

Fig. 2 : ^{15}N variations in organic matter of a north German peat bog

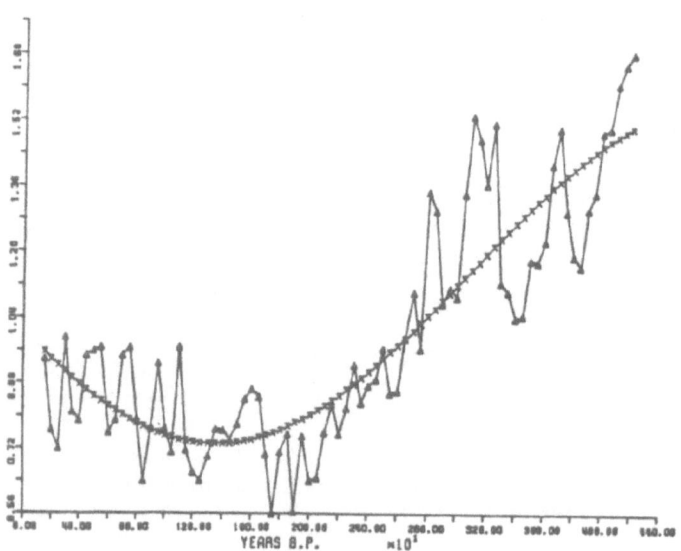

Fig. 3 : General long term trend (xxx) of organic nitrogen in a north
German peat bog

A Fourier Analysis of the δ^{15}N-data reveals a statistically significant periodicity of about 1025 yrs and two further, slightly less significant periodicities of about 280 and 170 yrs. While the first value turns out to be rather well defined the latter two seem to have varied somewhat with time.

Combining the 3 extracted periodicities results in a δ^{15}N curve as shown in Fig. 4, together with the experimental values. This plot indicates a rather good approximation by assuming that certain events reoccur at about 170, 280 and 1025 yrs. The somewhat less well matched part of the curve beyond 2200 yrs may be due to a slight shift in the two lower periodicities not accounted for in this representation.

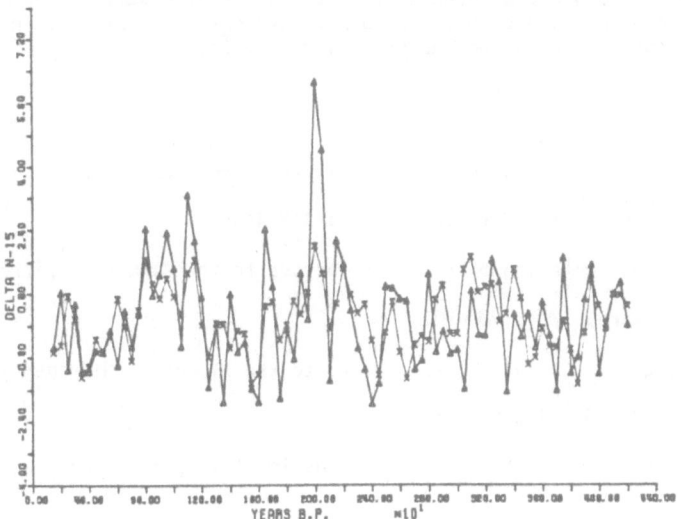

Fig. 4 : Record of δ^{15}N in a north German peat bog (▲▲) compared with the synthesis of 3 harmonics (170, 280 and 1025 years) which dominate the experimental curve (xx).

It is known that under certain conditions the degree of decomposition in peat bogs reflects past climatic variations [4]. However an interpretation i.e. correlation between certain climatic situations and the degree of decomposition is not straightforeward. For example increasing humidity by which the process of humidification slows down may either be attributable to higher precipitation or lower tempearture or even a combination of the two. Therefore a shift to colder or more humid weather conditions could produce the same humidification pattern (accumulation of more light coloured peat). In principle the same problem should be true for the corresponding δ^{15}N-values. But is is reasonable to assume that decay processes result in ^{15}N depleted nitrogen compounds by which the remaining organic matter is enriched in ^{15}N. With progressive humidification the corresponding organic matter therefore appears to increase in its ^{15}N-value. If this procedure represents the main mechanism one would expect for example the occurence of low ^{15}N-values from 1500-1700 because unfavourable conditions prevailed for this time period. The ^{15}N results agree well with the climatic development during this period, i.e. a beginning decline of temperature, which marked the onset of the "little ice age" (Fig. 2).

However a lot of questions still remain to be answered : f.e. what different plant communities have developed within various time periods and how would ^{15}N be affected tereby; what are the underlying fractionation processes of such a complex system, a.s.o.

Therefore more detailed interpretations necessitate further measurements to substantiate the above reasoning.

Nevertheless indications are at hand that peat bogs might enable statements of long term trends in climate which up to now have only been based on oxygen isotope measurements in ice cores and tree ring date [5, 6].

References

1 Senstius, M.W., 1925. The formation of soils in equatorial regions, with special reference to Java.
   Amer. Soil Survey Assoc. Rpt. Bul. 6, 1 : 149-161.

2 Jenny, H., 1928. Relation of climatic Factors to the Amount of Nitrogen in Soils.
   J. Amer. Soc. Agron. 20 : 900-912.

3 Jenny, H., 1929. Relation of Temperature to the Amount of Nitrogen in Soils.
   Soils Science 27, 3 : 169-188.

4 Aaby, B., 1976. Cyclic climatic variations in climate over the past 5500 yrs reflected in raised bogs.
   Nature 263 : 281-284.

5 Dansgaard, W., Johnsen, S.J., Reeh, N., Gundestrup, N., Clausen, H.B., Hammer, C.U., 1975. Climatic Changes, Norsemen and modern man.
   Nature 255 : 24-28.

6 La Marche Jr., V.C., 1974. Palaeoclimatic Inferences from Long Tree-Ring Records.
   Science 183 : 1043-1048.

CLIMATIC INDEXES ON THE BASIS OF SEDIMENTATION PARAMETERS IN GEOLOGICAL AND ARCHAEOLOGICAL SECTION

Roland PAEPE[1], Miranda Evangelia HATZIOTIS[2], Jacques THOREZ[3], Elfi VAN OVERLOOP[1] and Gaston DEMAREE[4]

The need for numerical parameters is a first requisite for mathematical evaluation of climatical evolution.

In continental sequences such parameters are seldom available. Whereas oxygen-isotope and foraminiferal investigations lead to quantitative climatical evaluation in deep-sea cores, such approach is hardly possible in continental deposits. With the exception of pollenanalytical and diatom quantification, continental series are hitherto exempt of any such treatment. Moreover continental sediment-series show a great complexity and heterogenety in composition unlike the regular sedimentation pattern of deep-sea cores.

Establishing accurate timescales, taking into account the stratigraphical hiatuses, is a most difficult entreprise for continental series. Further- more stratigraphic boundaries are of different nature as well : erosinal, pedological, diffuse, pebble bands, geochemical, etc. Hence, it is difficult to disentangle whether such boundaries are abrupt or continuous. Besides, abrupt lithostratigraphical changes do not infer necessarily with drastic climatic changes. At this point reading of macroscopic aspects of the lithostratigraphical sequence must be completed with laboratory investigation. As it is most often the case, in comple absence of bio- stratigraphical elements, sedimentological parameters are the only ones to be left over.

In the last fifty years numbers of sedimentological parameters have been worked out (KRUMBEIN, PETTIJOHN, McKEE) mainly on basis of simple grainsize analysis. The aim of such studies primarily was a better description of the sediment itself and if possible, a better understanding of the modalities of sedimentation (i.e. fluviatile, eolian, marine or lacustrine). A climatic bias was seldom the immediate goal. At the best conclusions as to the braided character of a river or to the eolian structure of a dune without specific indication of the climate dealt with, were drawn. The problem with this kind of approach is any absence of continuity since not all of the stratigraphic layers concerned are taken up for this type of investigation. DOUGLAS was perhaps the only one to consider continuous grain size variation in order to establish curves of different texture-mixing as a result of variable sedimentation energy rates.

Two examples are chosen to illustraite the possibilities in this field. The first one is dealing with climatic variations of the Weichselian (Last- Glacial) periglacial area; the second with Holocene climatic variations of the Eastern Mediterranean.

1 - Vrije Universiteit Brussel and Belgian Geological Survey, Belgium.
2 - Greek Archaeological Survey, Greece.
3 - Université de Liège, Belgium.
4 - Royal Meterological Institute, Belgium.

1. SEDIMENTOLOGICAL MATHEMATICAL APPROACH OF WEICHSELIAN PERIGLACIAL SERIES IN BELGIUM

Weichselian periglacial series of the Flemish Valley in Northern Belgium show an alternation of aggradation and standstills. The latter appear as sudden breaks i.e. erosion boundaries, accompanied or not with frost-wedges, pebble bands, vegetation or soil horizons.

The lithostratigraphical profile of Zemst (just North of Brussels) (Fig. a) have been tested on 14 grainsize analytical sample occurring in 17 layers above the pebble band with frost-wedges dated at about 55.000 yrs B.P. According to pollenanalytical studies (R. VANHOORNE, 1971) layer 11 may be evaluated in time at about 27.000 yrs B.P.

The grain size distribution of the 14 samples plotted on a Douglas-diagram show a gradual evolution, sample nr. 13 providing a perfect Gauss-distribution which plots as a straight line. This very same sample 13 occurring in layer 11 occupies an transitional position between the lower fluviatile and the upper eolian sedimentation series. This explains perfectly the straight line of sample 13 between the fading concavity of curves 1 through 8 and the increasing convexity of curves 9 through 14.

The gradual shift from fluviatile to eolian sedimentation curves is explained as being a result of gradual changes in the original water regime under influence of an increasing eolian activity. The latter is a direct consequence of steadily growing polar desert conditions. Actually as of layer 10 an increasing number of frot-wedge horizons is found.

Taking the median grain size values (Md) through time on semi-logarithmic paper one obtains two exponentional functions for both eolian and fluviatile series ($Md = k.e^t$).
It points to the fact that full eolian sedimentation conditions occurred as of sample nr. 8 whereas sample nr. 13 showing a Gauss-equilibrium was related to the final stage of the fluviatile sedimentation.

DEMAREE and PAEPE (1981) showed that the first set of data (1 through 8) are better represented by a linear relationship of the type $y = a_1 x + a_2$. However, differences with the exponential model are not significant at the 90 % confidence level (S. BRANDT, 1976).

The second set of data (8 through 14) is definitely better represented by an exponential curve of the type $y = a_1 e^{a_2 x}$.

Not linear regression techniques make it possible to determine the parameters estimates which minimize the sum of squares SSQ of the residuals having :

$$SSQ = \sum_{i=1}^{M} (y_i - f(x_i, a_i, \ldots, a_N))^2 \qquad (1)$$

with M being the number of observations x_i, y_i and f the non-linear function of the unknown parameters $a_1, \ldots, a_N$ ams of x_i, i = 1, ..., M.

A model function f which would assess the time of separation between eolian and fluviatile sedimentation periods was searched for. It should contain a parameter to separate the two curves while still allowing for a continuous evolution between both systems.

For statistical reasons (such as parsimony of parameters, small correlation between parameter estimates and statistical precision on the parameter estimates) the following model has been selected :

$$f = a_1 x_1 \cdot U (a_3 - x_i) / Ce^{a_2 x_i} U (x_i - a_3) \qquad (2)$$

The step function U cuts off the contributions to f which are not relevant to the segment under consideration.

U(v) is defined as follows :

$$U(v) = 0 \text{ if } v < 0$$
$$\tfrac{1}{2} \text{ if } v = 0$$
$$1 \text{ if } v > 0$$

For reasons of continuity between both parts of the graph, constant C is equation (2) takes the value

$$C = a_1 a_3 e^{-a_2 \, a_3}$$

As a results :

$$f = a_1 x_i \cdot U (a_3 - x_i) + a_1 a_3 e^{a_2 (x_i - a_3)} \cdot U (x_i - a_3) \qquad (3)$$

The 90 % confidence intervals, parameter estimates a_i and variation coefficients for individual parameter estimates possess all required statistical properties.

In conclusion, the evolution of median grain size values with time follows a linear (perhaps an exponential) law in the eolian sedimentation phase and a must faster exponential law in the fluviatile sedimentation phase. The transition between the two phases is gradual and its occurrence is estimated at 44.740 yrs B.P. ($\pm$ 1.294 yrs). This is at the beginning of the Middle Weichselian (or Last Glacial), i.e. the Moershoofd interstadial.

2. SEDIMENTOLOGICAL - MATHEMATICAL APPROACH OF HOLOCENE SERIES IN GREECE

Roland PAEPE, M.E. HATZIOTIS and E. VAN OVERLOOP established the archeogeological stratigraphy for the Haradros and Kifissos rivers respectively in Marathon and Western-Athens (Greece) for the Holocene i.e. the last 10.000 years.

As of 9.000 yrs B.P., twelve Holocene (Palaeo)soils (H.S.) occur at regular distances in time. They represent periods of standstills in between phases of diminantly fluviatile aggradation. Furthermore, their age could be satisfactorily determined in considering archaeological

levels at the top of each soils and the archaeological content of
fluviatile deposits in which the soils developed.

Most accurate were the age determination of the top-levels providing
tombs, constructions and occupation-levels with sherds, etc.

Its was found that each of the Holocene Soils are coinciding mainly with
the beginning of an archaelogical period. On H.S. 1 (Marathon Soil)
and H.S. 2, Neolithic tombs are found; on H.S. 3, H.S. 4 and H.S. 5,
tombs of the Proto-Helladic, Meso-Helladic and Hystero-Helladic (Mycenean)
Periods; on H.S. 6, H.S. 7 and H.S. 8 tombs and occupation layers of
respectively the Geometric, Classical and Hellinistic Periods.

Geologically speaking H.S. 1 and H.S. 2 correspond to the beginning of
respectively the Boreal (II) and Atlantic Sub-Stages (III); H.S. 3 and
H.S. 4 with the beginning and end of Sub-Boreal phase IVa; H.S. 5 with
the beginning of Sub-Boreal phase IVb. It points to an intemate relationship of the geological Stages and the archaeological periods. The
explanation is obviously related to possibilities of occupation of walley-
bottoms. Each soil-development phase allows the establishment of an occupation phase in the valleys which hitherto was impossible because of the
danger of floodings. One may not loose out of sight that during the first
half of the Holocen valleys were still deeply incised and narrow as a
result of the Last-Glacial incision due to a considerable drop in sea-
level.

About 700 B.C. i.e. approximately 2.700 yrs B.P. the former incised
valleys got completely filled up. At this level, the Kallikleios Soil
(H.S. 6) developed representing at the same time a benchmark in the
archaeological, geological, geomorphological and climatical evolution.
The soil development coincides indeed with the transition of the Sub-
Boreal (IVb) to Sub-Atlantic -Va). Culturally speaking one is entering
the famous periods of Archaic, Classical and Hellinistic times. From now
on, occupation of the valleys was not hampered by floodings anymore.
Actually floodings were localised to the very streamchannel and not so
much to the valley as a whole. The latter had become a broad plain where
ample space was available for occupation at all times.

General occupation of valley-plains continued during Roman, Byzantine,
Ottoman and Present Periods.

The difference with the aforegoing periods remains in the fact that soil-
development is related with subsequent phases of drought e.g. H.S. 6,
H.S. 9, H.S. 11, H.S. 12.

These droughts are known from ancient authors to occur during the 8th C.
B.C., 0 B.B., late 12th C. A.D. and of course today. Within the timespan
of the Sub-Atlantic, they occur at about a 1.000 years intervals. More
or less the same time-interval appears to exist for the development
phases of H.S. 5, H.S. 4 and H.S. 3 and even for H.S. 2 and H.S. 1.
Lack of evidence within the development phases of H.S. 2 and H.S. 3 does
not allow any conclusion for this timespan of three millenia covering
the Atlantic. New excavations although seem to give indications for
further subdivisions.

Anyhow, the general trend in soil development reveals a cycle of 1.000
years, coinciding with periods of drought as of 2.700 yrs B.P. i.e. the
Sub-Atlantic.

Each soil-development occurs at the end of a fluviatile cycle. Each fluviatile cycle shows a certain degree of aggradation rate. It becomes therefore quite feasible to interpret the phases of soil-development as phases of aggradation with zero-value.

In establishing the climato-sedimentation cycle of the Holocene Period on the basis of sedimentation-rate, one obtains the dashed-line curve. Here, the middle of the inter-period between soils corresponds to the maximum aggradation.

This "sedimentation curve" points after a moderate phase during the Boreal to a maximum during the Atlantic, and as steady weakening during the various Helladic Periods ending up with H.S. 6 i.e. the Kallikleios Soil.

New rises occur during the Archaic. Early Byzantine and Early Ottaman Periods.

The "sedimentation curve" thus reprensented, differs greatly from J. THOREZ' clay-weathering curve. Taking, however, the principle into account that soil-development occurs as standstill of zero-value at the end of each fluviatile phase, the "aggradation-peak" is shifted as a cumulative point to the position coinciding with each palaeosoil.

On the graph it is represented by the full-black curve showing following characteristics : a smoothly rising initial line from H.S. 1 (9.000 yrs B.P.), through H.S. 2 reaching its maximum at H.S. 3 (4.700 yrs B.P.) followed by a sudden drop to H.S. 4 and H.S. 5 reaching its maximum at H.S. 6 about 2.700 yrs B.P. In geological terms it shows a constant rise from Pre-Boreal through Boreal and Atlantic reaching its maximum at the beginning and a quick drop at the end of the Sub-Boreal.

Hereafter, this cycle repeats itself frequently in between phases of drought during the Sub-Atlantic corresponding to phases Va, Vb1 and Vb2.

The trend of the adapted sedimentation-rate curve in which the zero-values of the soil-stages are added to the shifted maxima of the aggradation, shows a remarkable resemblance to the clay-weathering curve. Terefore it is considered to be representative of the climatic evolution.

The next point to control is the relationship of the sedimentation-rates with the median grainsize-values and the pollenanalytical evolution.

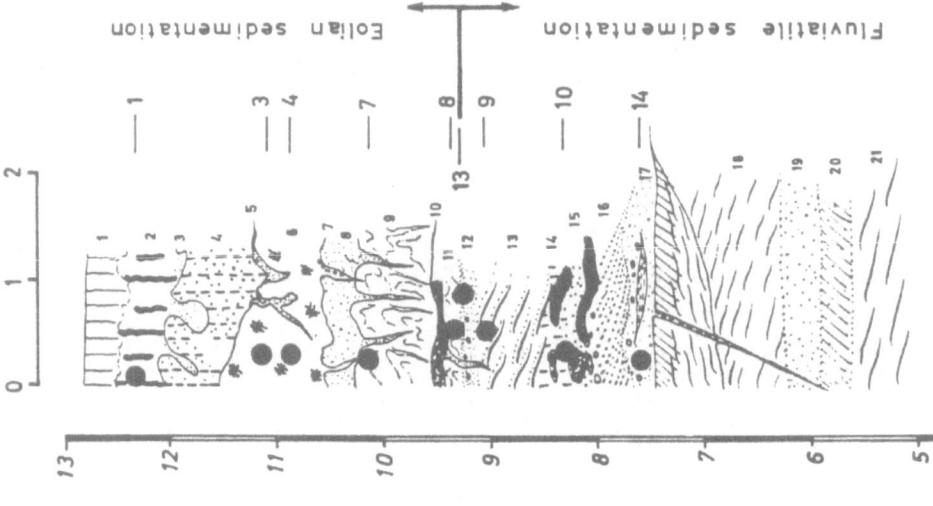

Profile section of Zemst.

1. Grey brown loam with A_p-horizon. 2. Mottled textural-B-horizon. 3. Yellowish sand and gravel layer; disturbed (frost wedges). 4. Yellowish brown sandy loam. 5. Large frost wedge row. 6. Mottled sand loam. 7. Frost wedge row with pebbles. 8. Green loamy sand. 9. Brown mottled fine stratified loam with cryoturbations; top limit with frost wedges. 10. Grey clayey loam with frost wedges at base 11. Greenish loamy sand. 12. Gravelly layer. 13. Grey sand with climbing ripple marks. 14. Disturbed loamy layer. 15. Grey white sand. 16. Coarse stratified sand. 17. Yellow fine sand. 18. Loamy sand with climbing ripple marks and frost wedges along upper limit. 19. Coarse sand. 20. Oxydized zone. 21. Loamy sand with climbing ripple marks.

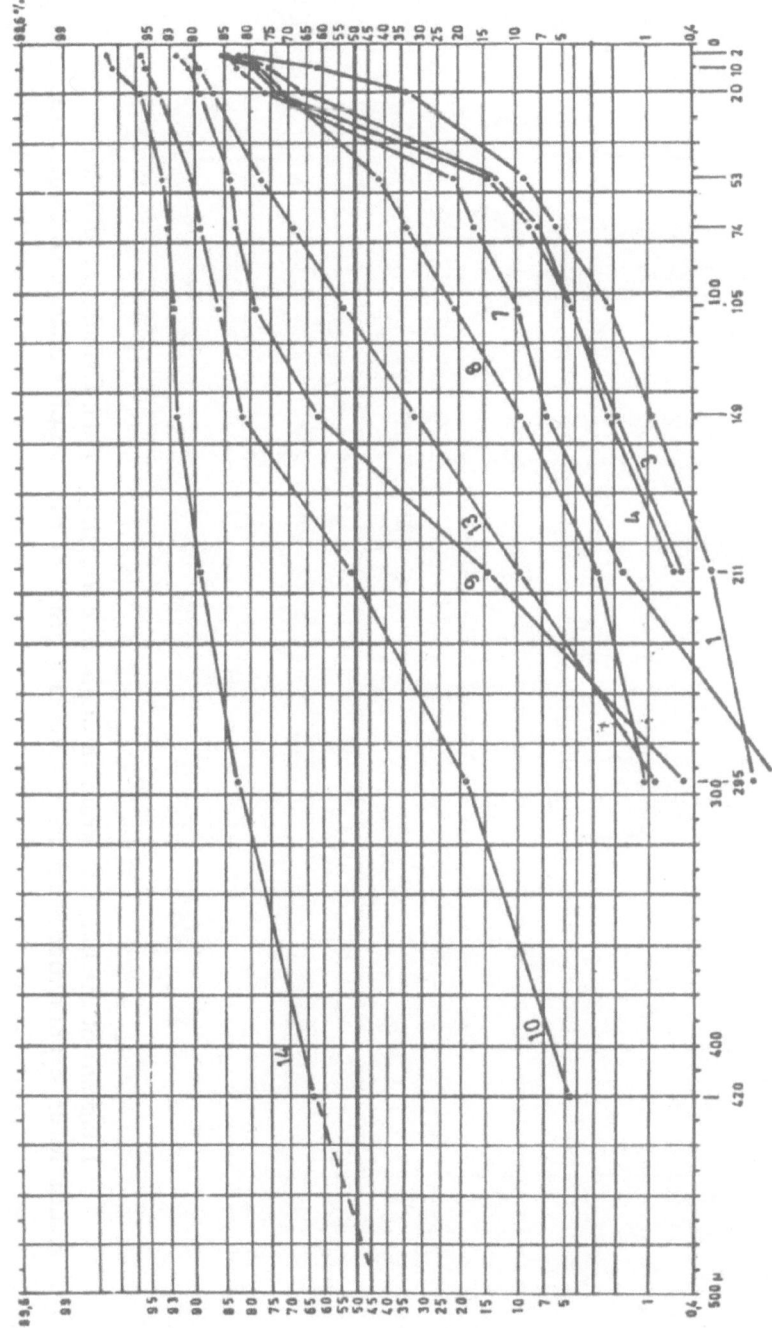

Grain size distributions plotted on Douglas diagram. On the latter diagrams, a Gauss frequency distribution plots as a straight line.

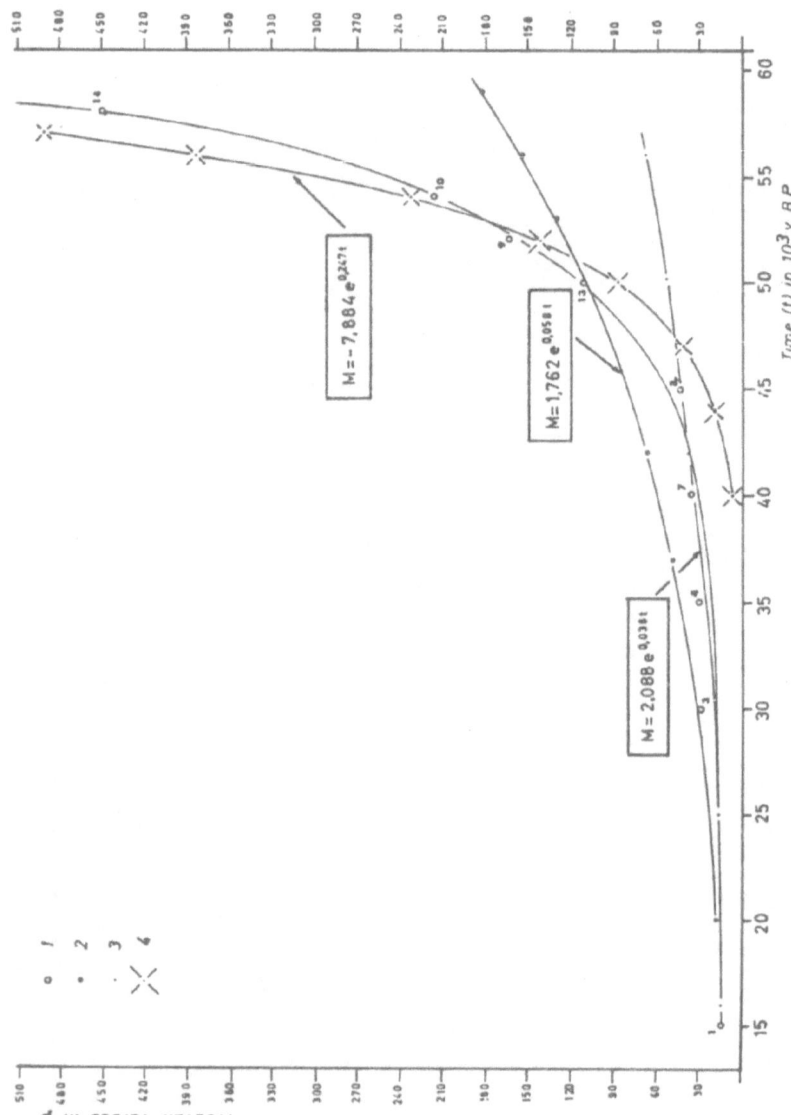

Diagram of median size grain through time (regular paper). Calculation of this curve results in an exponential function, thus giving proof of the complex change through time.

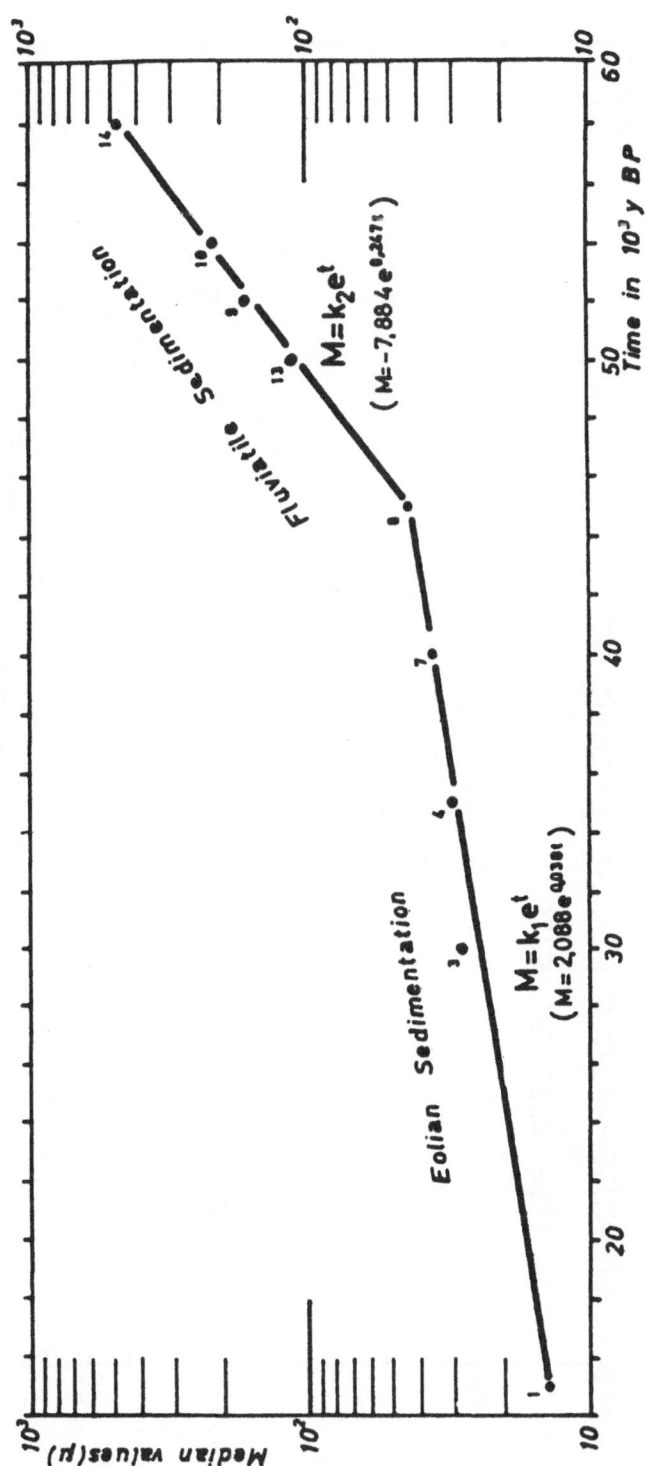

Diagram of median grain size through time (semi-logarithmic paper). The exponential function reads as a straight line on semi-logarithmic paper.

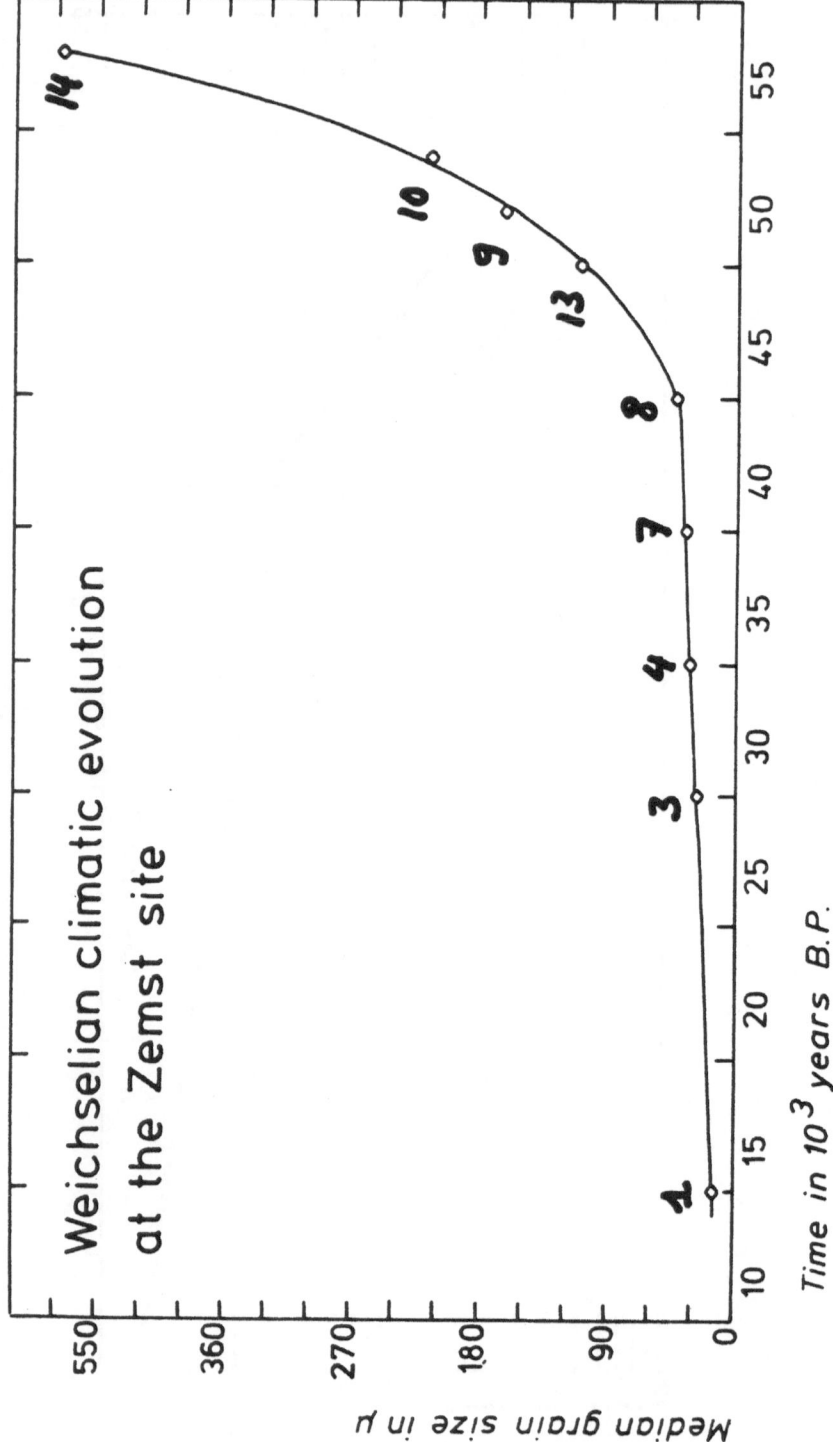

Weichselian climatic evolution
at the Zemst site

Median grain size in µ

Time in 10³ years B.P.

RATHER LONG DURATION OF THE TRANSIENT CLIMATIC EVENTS IN THE "GRANDE PILE"
(VOSGES – FRANCE)

G. SERET – Univ. of Louvain-la-Neuve (Belgium)

"Grande Pile" peat bog is located in the southwest slope of the Vosges
Mountain (France). It was formed by a glacial overdeepening in the
Buntsandstein sandstone due to an old Pleistocene ice-cap covering most of
the area. In its deeper part, the thickness of the "Grande Pile" lacustrine
sediments reaches about 30 m. The bottom is a cover of lodgement till
($\pm$ 5 m), surmounted by melt out till ($\pm$ 3 m) and lacustrine bottomset beds
($\pm$ 2 m) deposited within stagnant dead ice where the ice-cap receded.
Younger glaciers did not reach the area.

The last 20 meters of sediments are mainly of organic origin. The "Grande
Pile" stands isolated on a watershed plateau, it is therefore protected from
any kind of sediments supplied by running waters after the final melting of
glacier-ice. Most of the detritic minerals are sediments of clay- and silt-
sizes winnowed during the very cold phases. This is loess.

Pleistocene stratigraphy – G. WOILLARD (1975, 1978a, 1978b) provided a famous
pollen analysis for the post-till sediments in the "Grande Pile". She
designated pollen zones by numbers from 1 to 21. In a paper in collaboration
with W.G. MOOK (1982), she has collected fifteen ^{14}C datations going back to
69.500 years B.P. (stages 21 to 9). Correlations with deep sea ^{18}O stages
where inferred from Emiliani (1955) and Shackleton (1969). The pollen
diagrams of G. WOILLARD showed several important climatic changes.

Over the Eemian, she distinguished the warm Melisey I and II, the cold
St-Germain I and II and the cold Lanterne I, II and III.

Duration of a true event : a tentative calculation

A particularly sharp transient phase occurs in the cores between stages 6
and 7 of the pollen record, between the warm St-Germain I (13.88 m depth) and
the cold Melisey II (13.79 m depth). This indicates that a whole climat
change could take place on 9 cm of thickness. This is quite a rather short
duration for a whole climatic change (Fig. 1).

As indicated by pollen analysis, this climatic change is a true event, and
not only a crisis. Graminae and Pinus overages begin to increase 1 meter
before the end of St-Germain I, while Carpinus and Quercetum mixtum receded.
The vegetation cover shows a progressive and long modification, announcing
the occurrence of a cold phase (Fig. 1).

We could try to calculate the duration of the 9 cm of thickness of the
transient phase, if the mean rate of sedimentation were known. From ^{14}C
datations (WOILLARD and MOOK 1982) one obtains :

1. 69.800 yrs – 49.800 yrs = 20.000 yrs for 260 cm in the core between 12.8 m
   and 10.2 m depth.

   Mean duration for 9 cm : 693 years.

2. 49.800 yrs – 40.000 yrs = 9.800 yrs for 100 cm in the core between 10.2 m
   and 9.2 m depth.

   Mean duration for 9 cm : 882 years.

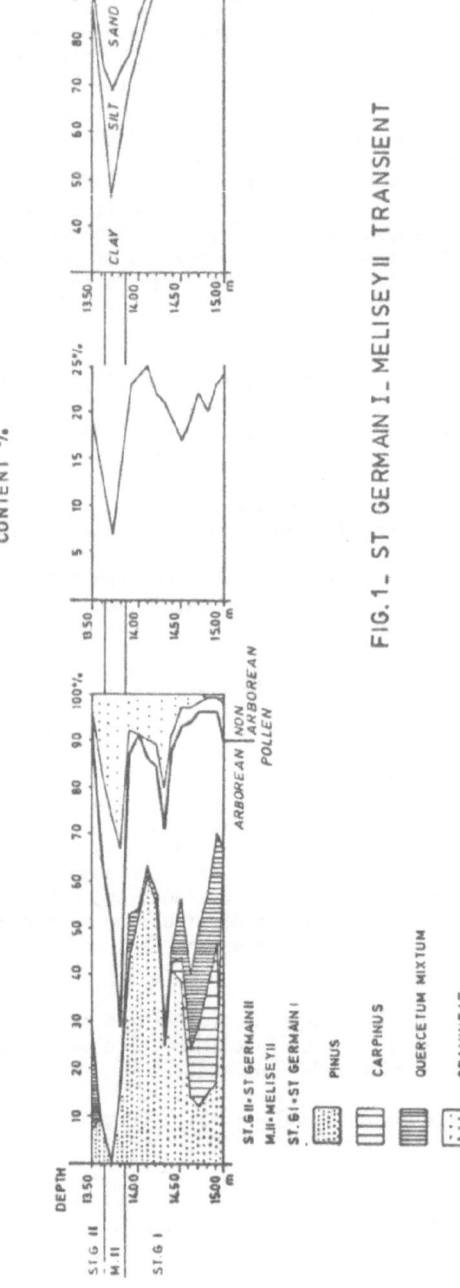

POLLEN DATA

ORGANIC CARBON CONTENT %

GRAIN SIZE ANALYSIS

FIG. 1 _ ST GERMAIN I _ MELISEYII TRANSIENT

3. 95.000 yrs - 73.000 yrs = 22.000 yrs for 180 cm in the core between 14.0 m and 12.2 m (from tentative correlation with Deep Sea ^{18}0 stages).

Mean duration for 9 cm : 1098 years.

The duration of the whole sharp transient phase would then be seven to eleven centuries.

Sedimentological data

Nevertheless, a duration of 7 to 11 centuries for the whole transient phase is surely too short, as indicated by sedimentology : organic carbon content, grain size analysis and sedimentary microstructures.

Organic carbon content fluctuates between 2.5 % for very cold phases and 33 % for the maximum warm periods. Palaeoclimatic correlations are very sharp with the pollen data curves. A rich organic carbon content in the warm periods is a normal consequence of a high biomass (Fig. 1).

Also grain size analisis correlates with palaeoclimatic data. Warm periods correspond to a clay content higher than 90 %, with very few silt and sand particles, whereas during the cold phases, the silt content becomes higher, up to 35 %. This is caused by the wind supply. The vegetation cover - mainly tress - recedes, particularly in the active alluvial plains, where proglacial and peripglacial deposits largely winnowed, due to the spreading of a more and more open steppe. A loess supply reaches the watershed plateaux as "Grande Pile" area. Loess appears in the core where simultaneous biomass largely decreases (Fig. 1).

It is important to note that loess supply began to appear in "Grande Pile" sediments before the local recession of the temperate forest species, as indicated by pollen records. This means that a change in vegetation took a long time to be completed the alluvial plains steppe adjoined temperate forest remnants on the plateaux.

Loessic sedimentation making its first appearance before the receding of the main AP curve in the core of "Grande Pile" indicates that periglacial conditions occurred before the general retreat of the trees on the plateaux. The transient climatic phases began then before the bending of the AP/NAD pollen curve. It began before the 9 cm measured on the curve and then lasted more than 7 to 11 centuries.

The study of thin sections in the basalm-indurated gyttja of the cores of "Grande Pile" indicates that the thickness of the layers is not at all proportional to the rate of sedimentation. At the depth of St-Germain I and Melisey II, between 13 and 14 meters, the material has been compacted. It presents very thin laminations of 10 to 20 micron thickness. It is possible to demonstrate that these laminations are true post-sedimentary structures. Figure 2 shows three 120 microns thick layers locally compacted by the pression of a sand grain, and presenting there 17 laminations. This pseudoschistosity has affected the softer layers, whose compaction has reduced the initial thickness of probably more than 50 %. A comparison with thin sections in silt rich layers does not show such post-sedimentary laminations (Fig. 3). Poor in organic matter, these layers were not soft enough to be affected by compaction. Their present thickness in the cores is very close to their primary thickness.

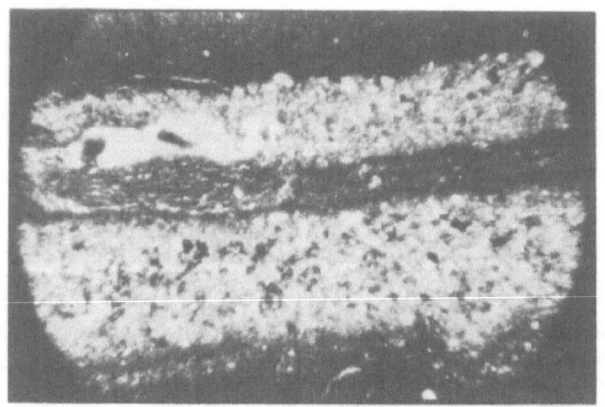

Fig.2. Post-sedimentary laminations in gyttja

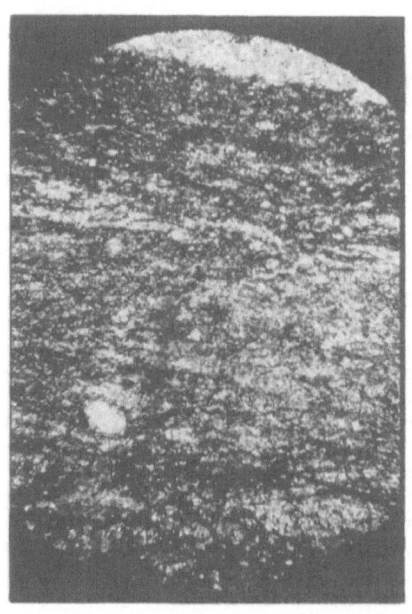

Fig.3. Lack of compaction structures
in mineral-rich layers.

Conclusion

The rate of sedimentation has changed during the filling of the "Grande Pile" lake. It was mainly consisting of organic matter deposited during the warm phases, and loess depodited during the cold phases. In addition, layers rich in organic matter were highly compacted. In the cores, theis present thickness at warm periods is proportionally less than that at cold phases. The existence of a correlation between mean thickness and mean duration is therefore dubious.

During a transient climatic phase, there is less biomass. A poor supply of loess poorly fed the bottom of the lake in sediments. The thinness of material of the transient phases doen not allow to considere short periods of time. So the St-Germain I - Melisey II transition lasted much more than seven or eleven centuries.

References

EMILIANI C., 1955. Pleistocen Temperatures. Journ. Geol. 63, pp. 538-578.

SERET G. 1967. Les systèmes glaciaires du bassin de la Moselle. Soc. Roy. Belge Géol., 577 p.

SERET G. and WOILLARD G. 1976. The glaciations in the Vosges Lorraines. In B. FRENZEL et al. - Führer zur Exkursionstagung des IGPC Projectes 73/1/24. Also in FRENZEL, Ed. (Bonn-Bad Godesberg 1978), pp. 1-30.

SHACKLETON N.J. 1969. The last Interglacial in the marine and terrestrial records. Proceedings of the Royal Soc. of London, B. 174, pp. 135-154.

WOILLARD G. 1975. Recherches palynologiques sur le Pleistocène dans l'Est de la Belgique et dans les Vosges Lorraines. Acta Geog. Lovaniensia, 14-1-L68.

WOILLARD G. 1978a. Grande Pile Peat Bog. A continuous Pollen record for the last 140.000 years. Quaternary Research, 9, pp. 1-21.

WOILLARD G. 1978b. The last Interglacial. Glacial cycle at Grande Pile in Northeastern France. Bull. Soc. belge de Géologie, 88. 1-51-69.

WOILLARD G. and MOOK W. 1982. Carbon^{-14} Dates at Grande Pile : correlations of Land and Sea chronologies. 215 - 159-161.

Session B : Initiation of Glaciation

THE OCEAN SURFACE DURING THE LAST INTERGLACIAL TO GLACIAL TRANSITION : A REVIEW OF THE AVAILABLE DATA

Claude PUJOL[1] and Jean-Claude DUPLESSY[2]

[1] Laboratoire de Géologie et Océanographie
Université de Bordeaux 1
Avenue des Facultés
33405 Talence - FRANCE

[2] Centre des Faibles Radioactivités
Laboratoire mixte CNRS-CEA
91190 Gif sur Yvette - FRANCE

The last interglacial was the last time that there was as small a volume of ice on earth as there is today. This period can be recognized in the benthic foraminiferal oxygen isotopic record (Fig. 1, from Dansgaard and Duplessy, 1981) as the last time that $\delta^{18}O$ values were as low as they are today. It coincides with isotopic substage 5e, using the nomenclature introduced by Emiliani (1955) and Shackleton (1969). This last interglaciation lasted only 11,000 years : it began about 127,000 years ago and finished 116,000 years ago.

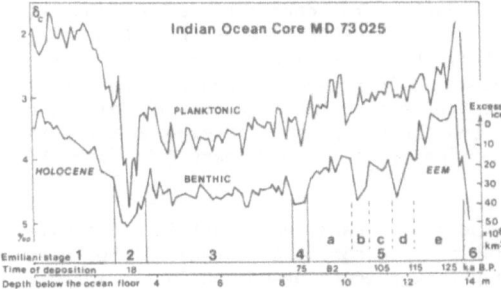

Fig. 1. Oxygen isotope δ, records of *planktonic* foraminifera, *Globigerina pachyderma* (upper curve), and of *benthic* foraminifera (lower curve) in the same core from the South Indian Ocean. The benthic record is assumed to reveal past continental ice volume in excess of the present, as indicated on the scale to the right (valid for the lower curve, only). The non-linear time scale is established by reconciling characteristic features (e.g. the Emiliani stages 2, 4 and 5b) with the corresponding features in other cores that have been dated independently. The figure suggests (1) that it is more than 100,000 years since the continental ice volume was as low as it has been in our own warm period, and (2) that at the end of the last warm period (Eem), the continental ice sheets and glaciers grew by some $40 \cdot 10^6$ km³, or 15 times the present volume of the Greenland ice sheet, within some 10,000 years.

Simulations of the Last Interglacial and of the last Interglacial-to-glacial transition with general circulation models are presently being made in relation with the Milankovitch theory of palaeoclimates. They simulate the primary response of the atmosphere to anomalies in the distribution of incident solar radiation by latitude and season introduced by the secular variations of the Earth's orbital elements. However other boundary conditions have to be specified and we review in this paper the available geological evidences describing the ocean surface during the initiation of the glaciation.

The study of this last interglacial was the last task of the CLIMAP Project. The final reconstruction is based on detailed oxygen isotopic analyses and biotic census counts of 52 cores across the world ocean (CLIMAP Project Members, 1983). This reconstruction shows that sea surface temperatures were only a little different than those of today, with weak evidence of a slightly warmer (1-2ºC) North Atlantic and North Pacific and a cooler Gulf of Mexico. By contrast, shallow marine and land sequences provide stronger evidence of warmer conditions during the last interglacial.   For example, in Europe, the last interglacia   is known as the Eem or Eemian interglacial   and many continental sediments deposited during this climatic episode have been found in Germany, Poland and Russia. Pollen analysis indicates that winter and summer temperatures were 2 to 4º C higher than those of today in Poland and Germany and even more in Russia (Frenzel, 1973).

Another part of the CLIMAP study focused on the relative sequencing of ice-growth versus oceanic cooling on the interglacial-glacial transition (transition between isotopic substages 5e and 5d). In the high latitudes of the southern hemisphere, the oceanic response preceded the global ice volume response (marked by the oxygen isotope signal) by several thousand years. Such a pattern can be observed in figure 1 in the planktonic foraminiferal isotopic record that exhibits a strong $\delta^{18}O$ increase in the middle of isotopic substage 5e. The reverse pattern is found in the subpolar North Atlantic Ocean from 40º to 60º N, which maintained warm sea surface temperature (even slightly warmer than those of today's ocean) during the first half of the ice-growth phase (Ruddiman and Mc Intyre, 1979).

In order to draw a map of the sea surface temperature in the world ocean at the middle   of the interglacial to glacial transition, about 116,000 years ago, we used the CLIMAP (1983) data and a few other cores from the North-eastern Atlantic studied by Pujol (1980). For each core, we obtained an estimate of the winter and summer sea surface temperature by taking the value calculated at the mid-point of the transition between isotopic substages 5e and 5d. Results are reported in figures 2a and 2b. We plotted (Fig. 3a and 3b) the winter and summer sea surface temperature difference between the last interglacial   and the mid-point of the 5e/5d transition by sub   the temperature estimates of 116,000 years B.P. from those of substage 5e.

Figures 2 and 3 show that the temperature change was small in the middle latitudes of the world oceans. The major diagnostic feature of the inception of the glaciation is the high temperature which prevailed in the North-eastern Atlantic Ocean late after ice began to pile up over the continents. A temperature drop of the order of 1 to 3º C already occurred in the Southern Ocean and in the North Pacific and in the Western North Atlantic. No informations on sea-ice changes during the 5e/5d transition are available at the present time, although they may be important in the Labrador Sea and in the Norwegian Sea.

BIBLIOGRAPHY

CLIMAP Project Members (1983)
        The Last Interglacial Ocean. Quaternary Res. (in press).

DANSGAARD, W. and DUPLESSY, J.C. (1981)
        The Eemian Interglacial and its termination. Boreas, 10, 219-228.

EMILIANI, C. (1955)
        Pleistocene temperatures, J. Geol., 63, 538-578.

FRENZEL, B. (1973)
        Climatic fluctuations of the ice age. The Press of the Case Western
        Reserve University, Cleveland.

RUDDIMAN, W.F. and McINTYRE, A. (1979)
        Warmth of the subpolar North Atlantic Ocean during northern
        hemisphere ice-sheet growth. Science, 204, 173-175.

SHACKLETON, N.J. (1969)
        The last interglacial in the marine and continental records.
        Proc. Roy. Soc. Lond., B, 174, 135-154.

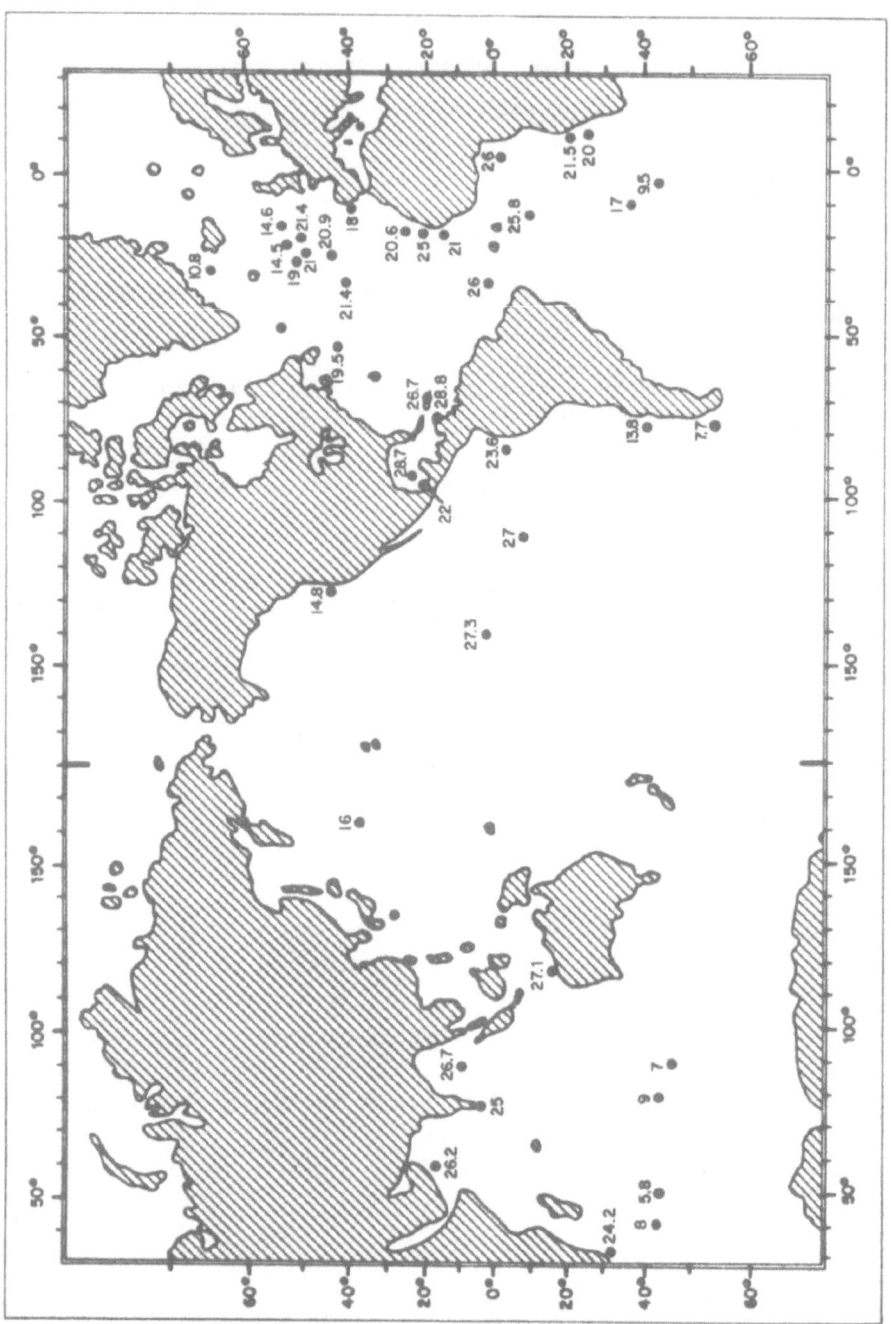

Fig. 2a : SEA SURFACE TEMPERATURE IN SUMMER (5e/5d transition)

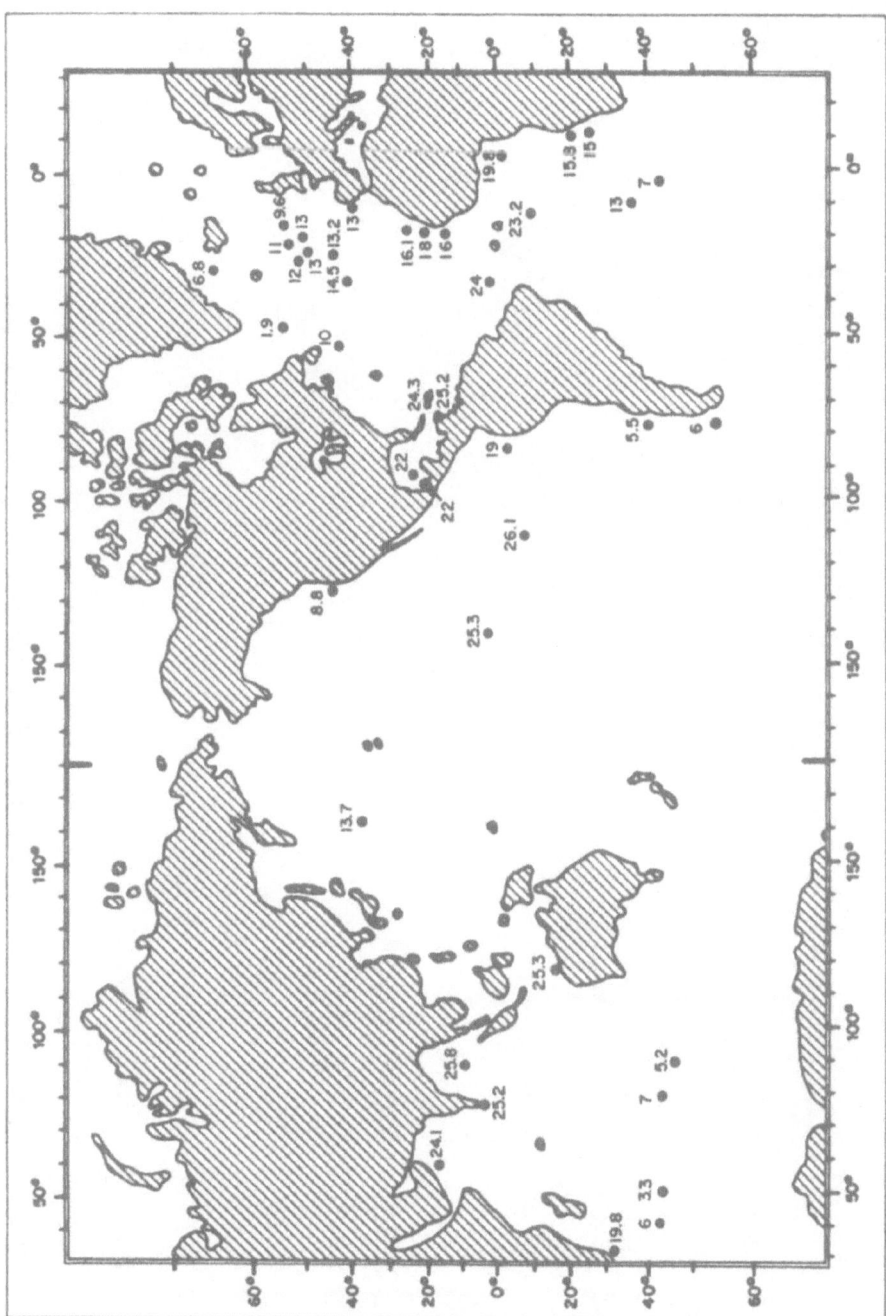

Fig. 2b. SEA SURFACE TEMPERATURE IN WINTER, 5e/5d TRANSITION

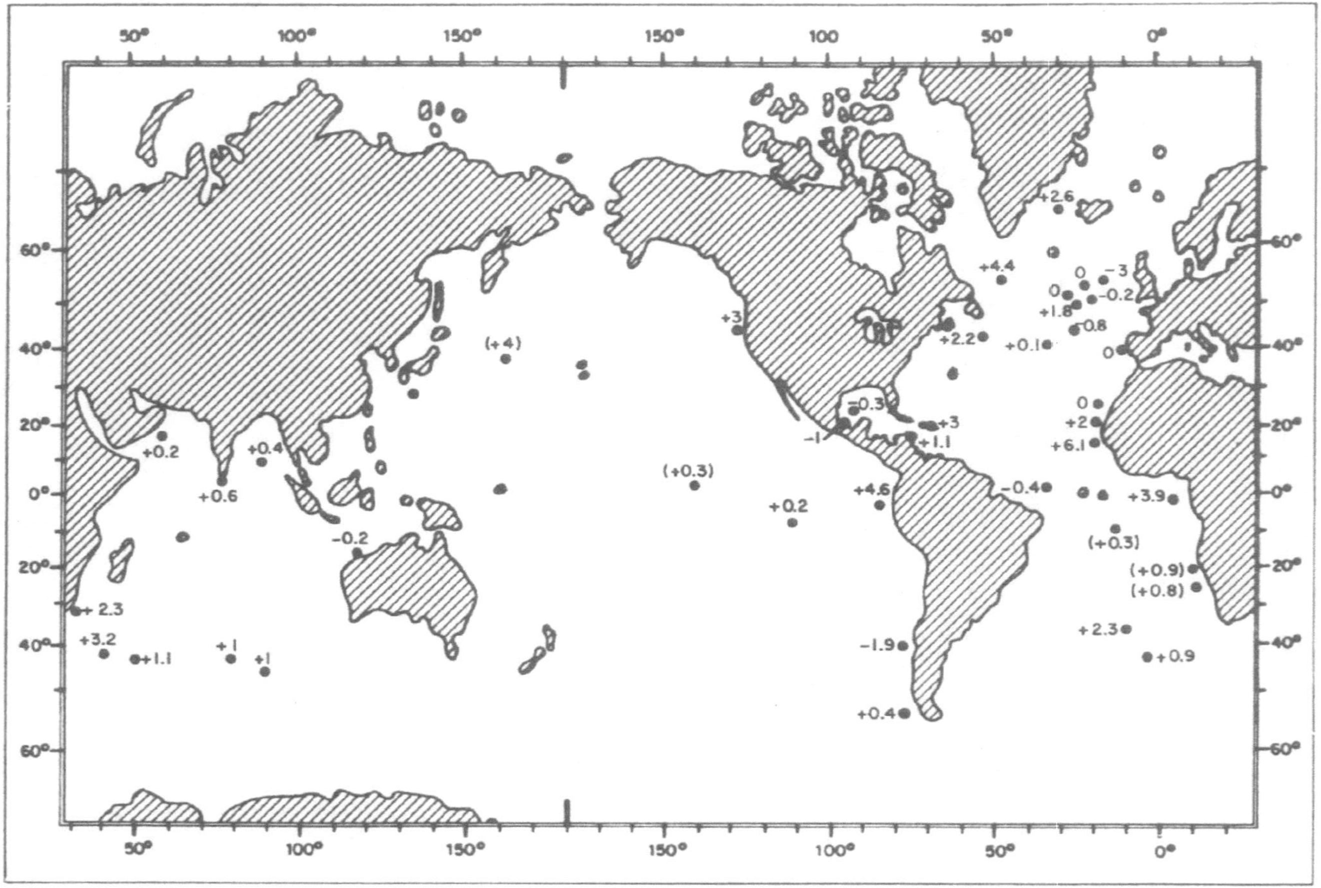

Fig. 3a. SUMMER SEA SURFACE TEMPERATURE DIFFERENCE BETWEEN 5e AND 5e/5d TRANSITION

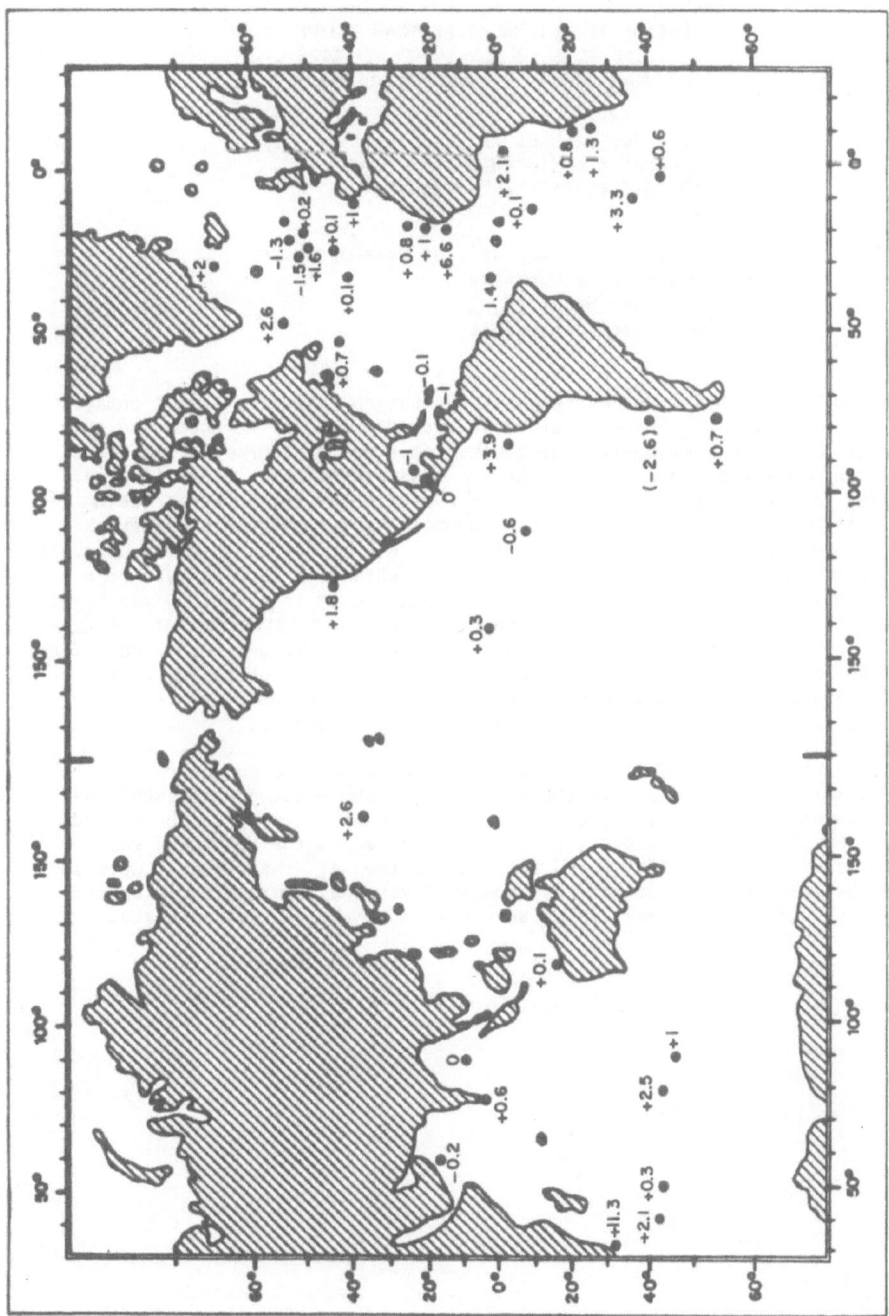

Fig. 3b. WINTER SEA SURFACE TEMPERATURE DIFFERENCE BETWEEN 5e AND 5e/5d TRANSITION

ABRUPT CLIMATIC EVENTS DURING THE LAST
GLACIAL TO INTERGLACIAL TRANSITION

Jean-Claude DUPLESSY[1] and Claude PUJOL[2]

[1]Centre des Faibles Radioactivités
Laboratoire mixte CNRS-CEA
91190 Gif-sur-Yvette - FRANCE

[2]Laboratoire de Géologie et Océanographie
Université de Bordeaux 1
Avenue des Facultés
33405 Talence - FRANCE

During the last glacial to interglacial transition, abrupt climatic changes
have been observed in the palaeoclimatic record, both in the ocean and over
the continent. We previously made a detailed study of four deep sea cores
with high sedimentation rate from the Northeastern Atlantic Ocean (Fig. 1)
in order to correlate the palaeoclimatic evolution of this oceanic basin
with that of the adjacent europen continent. The stratigraphic framework
was established by oxygen isotope analysis of planktonic/benthic foraminifera
and Carbon-14 dating. Sea surface temperature estimates were derived from
microfaunal analysis of sediments and pollen were used to establish vegetational pattern changes of Western Europe. We have demonstrated that the major
oceanic cooling that we recorded arond 10,400 Before Present (B.P.) coincides
with the continental Younger Dryas cold event, while the preceding warm
oceanic event correlates with the combined continental Bölling/Alleröd warm
episodes (Duplessy et al, Palaeo-3, 35, 1981, pp. 121-144).

The first phase of the deglaciation, which started at the end of the last
glacial maximum and ended with the beginning of the Bölling warm event, was
called Termination I_A. The second step of the deglaciation, which started
at the end of the Younger Dryas cold event and led to the Holocene, was
called Termination I_B (Fig. 2 and 3). Termination I_A, the climatic deterioration between the Alleröd and the Younger Dryas and Termination I_B consti-
tute three climatic changes, which were abrupt at the geological scale.

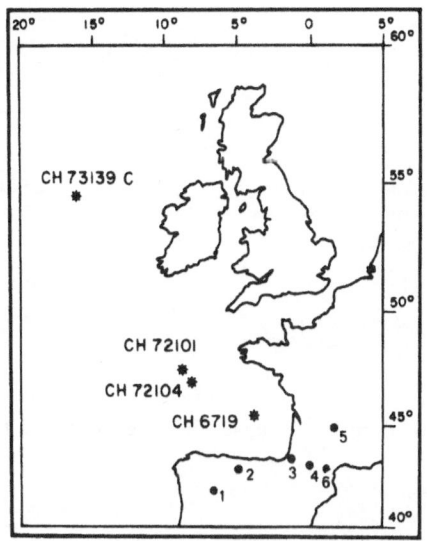

Fig. 1 - Location of core CH 73139C in the northeastern Atlantic Ocean and cores CH 6719, CH 72101 and CH 72104 in the Bay of Biscay. Numbers 1-6 refer to continental sites with palynological profiles;
1 = Lagunas de las Sanguijuelas
2 = Puerto de Rio Frio
3 = Le Moura
4 = Poueyferré
5 = Flageolet II
6 = Balcère

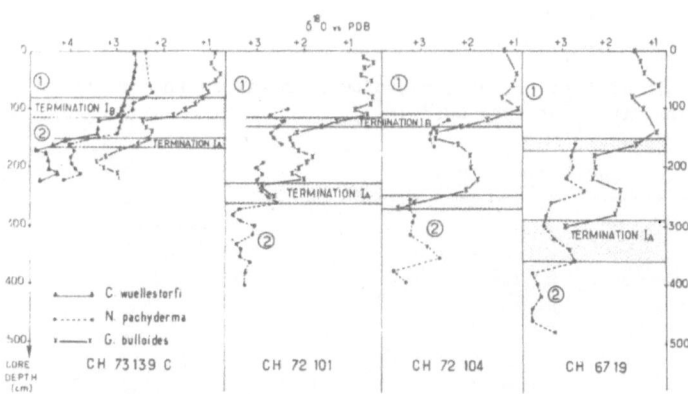

Fig. 2 - Oxygen isotopic records of CH 73139C, CH 72101, CH 72104 and CH 6719. Termination I_A and termination I_B represent the two major shifts of Broecker and Van Donk's termination I.

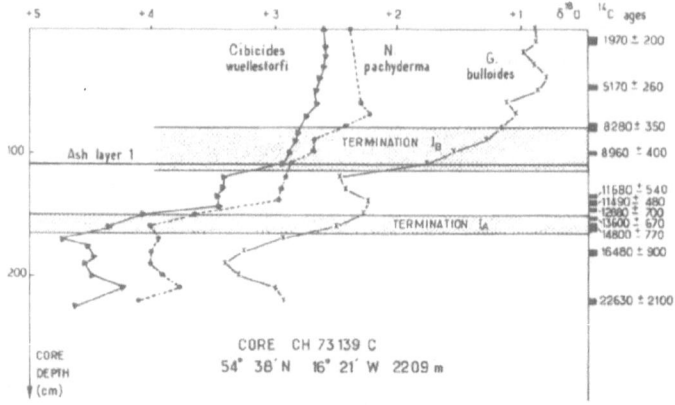

Fig. 3 - Oxygen isotopic record ^{14}C ages in core CH 73139 C

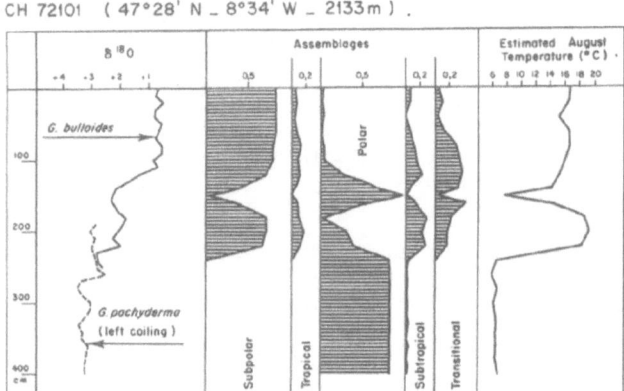

Fig. 4 - Oxygen isotopic record, microfaunal analysis and estimated August temperature for core CH 72101.

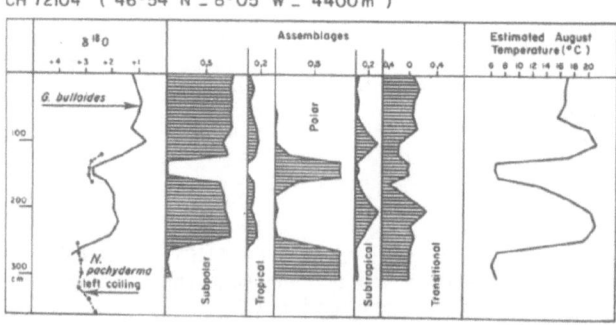

Fig. 5 - Oxygen isotopic record, microfaunal analysis and estimated August temperature for core CH 72104.

CH 6719 (45°44' N _ 3°57' W _ 1800 m)

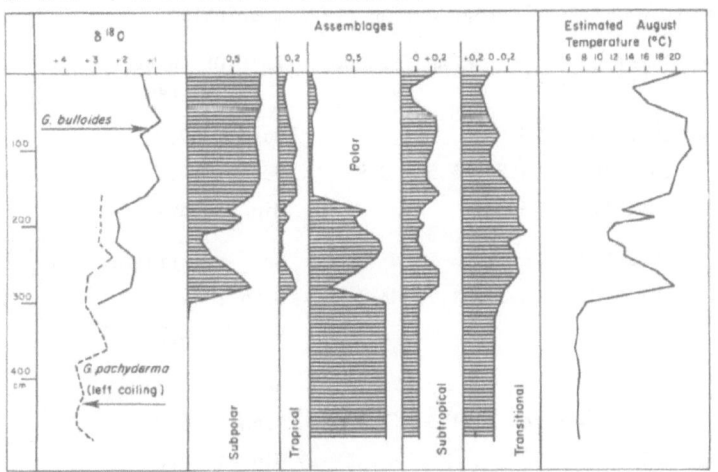

Fig. 6 - Oxygen isotopic record, microfaunal analysis and estimated August
temperature for core CH 6719.

By comparing the well dated isotopic record of core CH73-139C with the isotopic and palaeotemperature records of core CH72-101, CH72-104 and CH67-19
(Fig. 4, 5, 6), we estimated the rate of temperature change for surface water
in the Bay of Biscay during these transitions. Results are reported in
Tables 1 and 2.

TABLE 1

Climatic parameters and duration of the
major events of Termination I

EVENT	BEGINNING year B.P.	END year B.P.	DURATION Years
TERMINATION I$_A$	15,500 ± 800	13,300 ± 700	2200 ± 1500
Alleröd/Y. Dryas	11,000 ± 600	10,200 ± 500	800 ± 2200
TERMINATION I$_B$	10,000 ± 500	8,350 ± 350	1700 ± 850

TABLE 2

Rate of temperature change during the Deglaciation

EVENT	T CHANGE (°C)	MEAN T CHANGE (°C)	RATE OF CHANGE (° C/10^3 years)
	CH72-101 : 13		Max : 20.0
TERMINATION I$_A$	CH72-104 : 14,5	13.5 ± 0.9	Mean : 6.1
	CH67-19 : 13		Min : 3.4
	CH72-101 : 8		Max : infinite
Alleröd/Y. Dryas	CH72-104 : 14.5	11.5 ± 3.3	Mean : 14.4
	CH67-19 : 12		Min : 4.3
	CH72-101 : 10		Max : 16.0
TERMINATION I$_B$	CH72-104 : 14	11.3 ± 2.3	Mean : 6.7
	CH67-19 : 10		Min : 3.5

The major diagnostic feature of these results is the large uncertainty in the rate of temperature change; minimum values are of about 0.4° C per century, but maximum values larger than 2° C per century are also compatible with the experimental results.

The causes of this uncertainty are :

1. Bioturbation which mixes, more or less completely, the upper centimeters of sediment. However, our results show that, even the taking into account the record of core CH67-19, which is more bioturbated and disturbed than the two others cores, the uncertainty in the temperature variation estimates rarely exceeds 20 %.

2. Dating uncertainties due to statistical error in the radioactive counting. This error is high because foraminifera are not abundant enough in marine cores to furnish the large volume of CO_2, which is necessary for accurate carbon-14 counting.

As a consequence, the duration of the abrupt climatic changes recorded in deep sea cores during the last glacial to interglacial transition is known with an accuracy, which is not better than 50 to 100 %. More precise estimetes could be obtained in near future by the new technique of carbon-14 counting using an accelerator.

ICE-SHEET MODELLING FOR CLIMATE STUDIES

J. Oerlemans
Institute of Meteorology and Oceanography
University of Utrecht

1. Introduction

In recent years climatologists have become more and more aware of the
fact that the dynamics of large continental ice sheets may be a very
important factor in the evolution of climate. Apart form continental
drift, ice sheets probably possess the largest time scale in the climate
system. They effect the climatic regime through the well-known ice-
albedo feedback koop as well as by dynamical processes (e.g. production
of meltwater stabilizing the stratification in the oceans, forcing of
planetary waves in the atmosphere).

Even without knowledge of the physical principles governing ice flow, one
can easily understand why ice sheets are sensitive to changes in
environmental conditions. A typical height-to-width ratio of continental
ice sheets in 10^{-3}, implying that an ice sheet of "ice-age size" is a few
kilometers thick. So when an ice sheet builds up, its mean surface
elevation increases considerably. Noting that the mass balance (annual
ice-accumulation rate at the surface) generally increases strongly with
height, it is obvious that an ice sheet becoming thicker will expand.

The point is illustrated in Figure 1, where a north-south cross section
of a northern hemisphere ice sheet is considered. The schematic picture
is typical for conditions on the northern part of the American continent.
The climatic state is represented by a snow line, defined as the fictive
surface elevation at which the mass balance is zero. The intersection of
the snow line with sea level is called the climate point P, the intersection of snow line and ice surface the equilibrium point E. The
picture clearly shows that a small drop of the snow line, i.e. a small
southward shift of the climate point, may lead to a large ice sheet. Due
to the height-mass balance feedback the equilibrium point, which
determines the evolution of the ice sheet, can shift far southwards.

The ice sheets of Greenland and Antarctica are more restricted in their
response to changing environmental conditions, because these sheets are

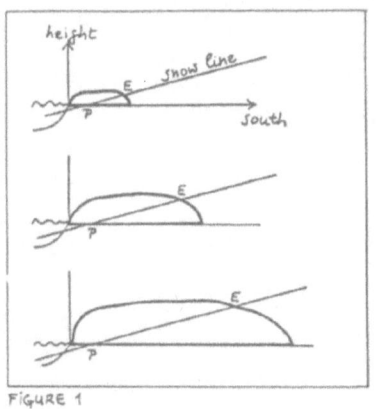

FIGURE 1

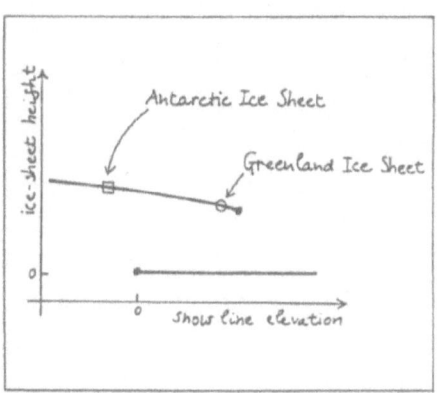

FIGURE 2

bounded by deep ocean. A typical steady-state diagram for bounded ice
sheets is shown in Figure 2. The continent is assumed to be flat with
zero surface elevation. It is obvious that for a snow-line elevation
larger than zero no ice sheet represents a steady state. For a moderately
high snow line, however, an ice sheet may be stable as well. The
consequent hysteresis appearing in the figure is a direct result of the
height-mass balance feedback. In the diagram the Greenland and
Antarctic Ice Sheets can be located. When the Greenland Ice Sheet would
be rmoved, it would not come back under the present climatic conditions.
The Antarctic Ice Sheet is subject to much colder conditions and therefore much more stable.

2. Modelling ice sheets

The flow of ice is governed by the following physical laws :

(i)   conservation of ice mass,

(ii) momentum equation,

(iii) thermodynamic equation,

(iv) a constitutive equation.

For present purposes ice can be considered as incompressible and of
constant density, so (i) reduces to zero divergence of the ice velocity
vector. In ice sheets accelerations are extremely small, implying that
(ii) is nothing but a balance between body forces (due to gravity) and
stresses. In the heat budget of an ice sheet, conduction, advection of
heat by the ice flow and dissipative heating are all important. The
constitutive equation (iv), finally, describes haw the ice deforms when
subject to forces. The simplified form of the set (i)-(iv) can only be
solved when proper boundary conditions can be formulated. This appears
to be a major problem in ice-sheet modelling. The treatment of the
bedrock-ice interface is particularly difficult.

Although calculating ice flow is a 3-dimensional problem, a useful simplification is to consider the vertically-integrated flow. Disregarding thermodynamics for the moment, the following equations give a reasonable description :

(1) continuity $$\frac{\partial H}{\partial t} = - \nabla . H\vec{u} + M$$

(2) stress equilibrium $$\vec{\tau}_b = \rho g H \nabla (H+h)$$

(3) constitutive eq. $$\vec{u} = A\tau_b^{n-1} \vec{\tau}_b$$

Here H is ice thickness, t time, $\vec{u}$ vertical mean horizontal ice velocity, M mass balance, $\vec{\tau}_b$ basal shear stress, ρ ice density, g acceleration of gravity, h bedrock elevation, A and n flow parameters, and $\nabla = [\partial/\partial x, \partial/\partial y]$. A typical value for n is 3, while A depends on tempe-rature.

Eq. (1) simply states that the time rate of change of ice thickness equals the divergence of the ice-mass flux plus the annual ice-accumulation rate. Eq. (2) gives the balance between horizontal pressure gradient force, acting on an ice column, and the shear stress at the base. Note that $\nabla(H+h)$ is the slope of the ice-sheer surface. Eq. (2) describes how the ice flows. It is based on the observation that ice deformation increases progressively with increasing forces (Glen's law).

This model can be tested by applying it to the Antarctic Ice Sheet, for example. Ice thickness variations of the Antarctic Ice Sheer are essentially due to the greatly varying bedrock topography and the variations in the mass balance (from 0.6 m ice depth per year in some coastel regions to less than 0.05 m ice depth per year in the central parts of the continent). To see how well (1)-(3) are able to reproduce the observed ice thickness variations a calculation was made on a 100x100 km grid. Replacing derivatives by finite differences eqs. (1)-(3) can be solved on this grid. With bedrock topography, edge of the ice sheet, and mass balance as input, the model can be integrated in time until a steady state is reached. The result is shown in Figure 3. Apparently, the simulation is quite reasonable, indicating that the approach described above contains the essential mechanism leading to the ice-thickness variations.

To increase the internal freedom of this type of model, the mass balance should be made a function of surface elevation, distance to open water (being the moisture source), sea-level temperature, surface slope. The latter is particularly important because it enables the model to let the orographically induced precipitation shift with the ice-sheet edge. Also, the ice sheet should be free to shrink or expand. When the mass balance near the ice-sheet edge is strongly negative, this gives no problems. However, when it runs into a shallow sea to form an ice shelf (floating ice, typically a few hundreds of meters thick), the ice sheet-ice shelf junction has to be modelled in order to keep track of the grounding line. Although several attempts have been made, a general procedure to do this in a reliable was has not yet been devised.

Fig. 3

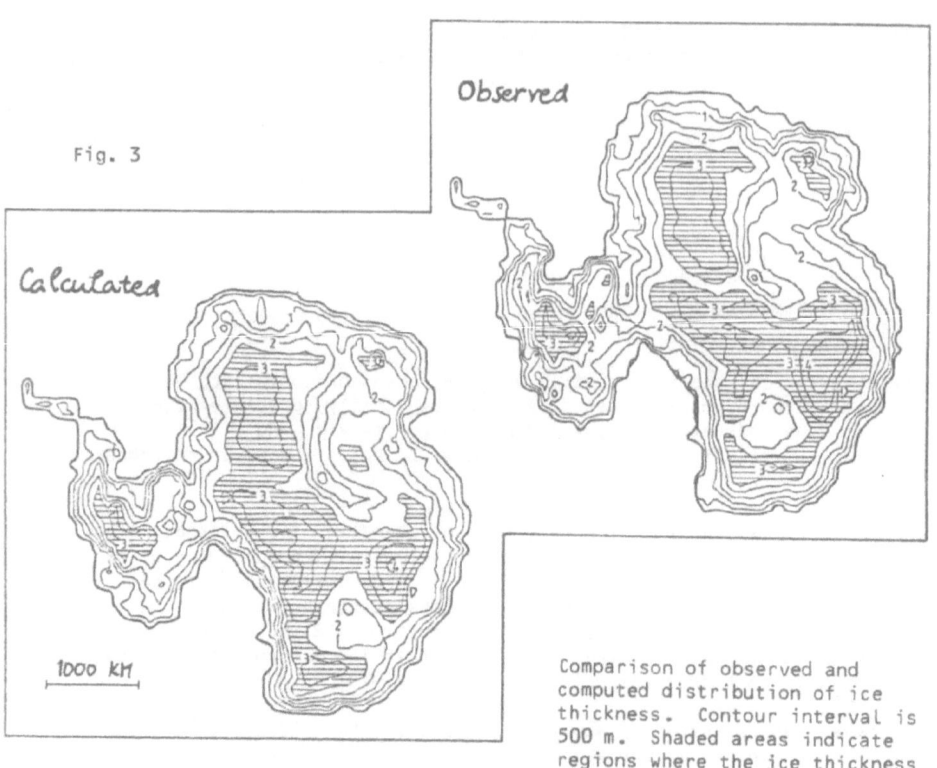

Observed

Calculated

1000 KM

Comparison of observed and
computed distribution of ice
thickness. Contour interval is
500 m. Shaded areas indicate
regions where the ice thickness
exceeds 2500 m.

3. Reaction of the bedrock to an ice load

The bedrock will of course react to variations in the ice load. For
modelling purposes, the lithosphere can be assumed to behave as a thin
elastic shell, resting on a viscous asthenosphere. For ice loads with
a large horizontal scale (> 200 km say), the rigidity of the lithosphere
can be neglected and the equilibrium state is one of isostatic balance :

$$(4) \qquad b = - \frac{\rho_{ice}}{\rho_{rock}} H$$

Here b is the bedrock height with respect to the ni ice load case.

To achieve this isostatic equilibrium material has to be displaced in the astenisphere. This takes time, depending on the viscosity of the astenosphere. In a first approximation one can simply assume that there is a local damped return to isostatic equilibrium on a uniform time scale T, i.e.

(5)
$$\frac{\partial b}{\partial t} = -[b + \frac{\rho_{ice}}{\rho_{rock}} H]/T$$

However, a more careful analysis of the flow in the astenosphere shows that T sould increase with the characteristic horizontal scale of the load. This leads to an equation of the type

(6)
$$\frac{\partial b}{\partial t} = K \frac{\partial^2}{\partial x^2} \left[b + \frac{\rho_{ice}}{\rho_{rock}} H \right]$$

where K is a constant depending on the viscosity of the asthenosphere.

Near the edge of the ice sheet the horizontal scale of the load is so small that here the rigidity of the lithosphere becomes important. This leads to the so-called forebulge, as shown in Figure 4. It depends on the specific application whether such effects have to be taken into accout. For example, when on studies the subtle grounding-line dynamics of an ice sheet in a shallow sea, lithospheric rigidity should be dealt with.

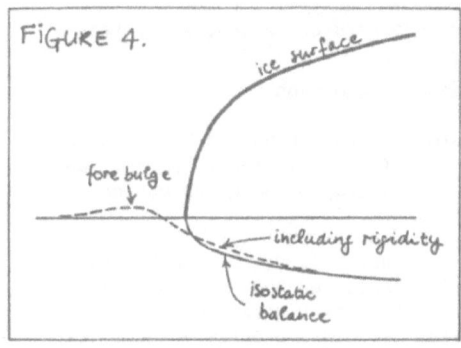

FIGURE 4.

4. Including thermodynamics

As noted earlier, the flow parameter A increases with ice temperature. Due to accumulation of geothermal heat, temperature in the lower layers of an ice sheet gradually increases until a balance between the upward geothermal heat flux and downward advection of cold ice is reached. For thick ice sheets the base may become subject to melting and basal sliding may occur. This substantially enhances the ice-mass discharge, and thus the evolution of the ice sheet. To deal with such effects in a numerical ice-sheet model, it is obviously necessary to calculate the temperature field in the ice sheet, i.e. to solve the thermodynamic equation.

This equation reads

(7)
$$\frac{\partial \Theta}{\partial t} = - \vec{v} \cdot \nabla \Theta + \mu \nabla^2 \Theta + Q$$

Ice temperature is denoted by Θ, the 3-dimensional ice velocity vector by $\vec{v}$, thermal conductivity by μ, and frictional heating by Q.

A complete solution of (7) requires the use of a 3-dimensional grid, which however leads to computational times that are much too large. Also, this would not be compatible with the relatively simple flow model discussed above.

A simplification that seems reasonable at this point is to expand the vertical temperature profile in a power series and to retain only a few terms :

(8) $\Theta(\vec{r},t) = \Theta_o(\vec{r},t) + z \Theta_1(\vec{r},t) + z^2 \Theta_2(\vec{r},t).$

Here z is height above the bedrock, and $\vec{r}$ the 2-dimensional location vector. So the temperature profile is constrained to be a second-order polynomal.

Three equations are now needed to calculate Θ_o (the basal temperature), Θ_1 and Θ_2. These can be derived from :

(i)   the upper boundary condition [the ice temperature equals the annual air temperature],

(ii)   the lower boundary condition [the temperature gradient a the base matches the geothermal heat flux],

(iii) the vertically-integrated thermodynamic equation.

In addition to this the shape of the velocity profile has to be prescribed because of the advective therm appearing in (7). Frictional heating can be calculated from the release of potential energy by downward motion.

As soon as basal melt water is produced, and both model calculations and observations suggest that for a large ice sheet this is rule rather than exception, the ice-mass discharge will increase. By how much is still an open question, and the same applies to how the basal water spreads. This problem will probably turn out to the most difficult problem in numerical ice-sheet models.

An example of the type of behaviour that can be caused by the effect of basal water is shown in Figure 5. An ice-sheet model including bedrock dynamics and thermodynamics was integrated in time for a fixed snow line (much lower than the present snow line !). The geometry is that of Figure 1. After 59 000 yrs basal water forms at the centre of the ice shett and slowly creeps southward (according to the parameterization of basal-water flow employed in the model, which is rather ambiguous, however). At t=68 000 yrs the water reaches the edge and a "surge" takes place. The sudden lowering of the ice surface, and the fact that the bedrock is still suppressed, causes raid melting of the southern half of the sheet. Then the ice sheet builds up again. So the effect of basal melt water may lead to oscillatory behaviour. It should be stressed, however, that such findings critically depend on the specific parameter-ization of basal-water flow and the effect on the ice-mass discharge.

Yet this experiment shows that bedrock sinking and production of basal water may bestabilize a large continental ice sheet. More detailed studies with this type of model concerning the origin of the pleistocene glacial cycles in fact show that without bedrock sinking and basal sliding the ice sheets, once they are there, do not disappear anymore.

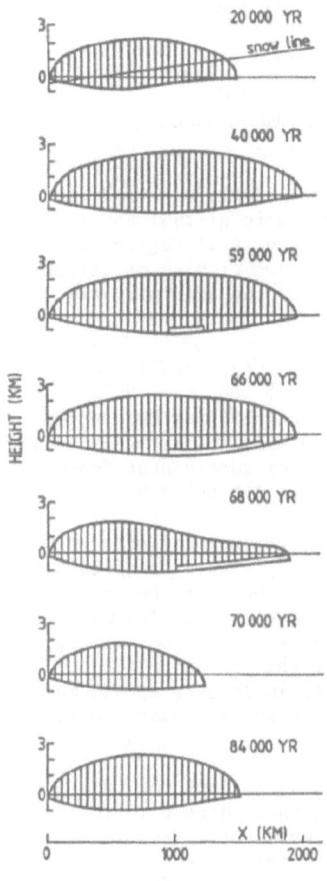

FIGURE 5.

Due to the interaction of ice flow and temperature field, the response of ice sheets to varying environmental conditions is complicated. Its study therefore requires careful numerical modelling. At this point the bottleneck seems to be the treatment of basal water. A proper treatment of the grounding line poses another problem. In studying the stability of the West Antarctic Ice Sheet, several schemes have been proposed to calculate the movement of the grounding line. However, so far only one-dimensional models have been constructed shich treat the ice sheet-ice shelf junction along a flow line.

5. Further reading

An introductory textbook on glaciology, dealing with both observational and modelling aspects, is : W.S.B. Paterson (1981), The Physics of Glaciers, Pergamon Press (second edition). As useful discussion on glacier modelling, in which a more fundamental treatment of ice-flow mechanics can be found, is : W.E. Budd and D. Jenssen (1975), "Numerical modelling of glacier systems", in Proceedings of the Snow- and Ice Symposium Moscow 1971 : IAHS-publ. 104, 257-291.

Experiments carried out by the author with the type of models described here can be found in : J. Oerlemans (1982), "A model of the Antarctic Ice Sheet", Nature 297, 550-553; and "Glacial cycles and ice-sheer modelling", in Climatic Change 4, 353-374.

A G.C.M. SIMULATION OF THE IMPORTANCE OF INSOLATION FORCING FOR THE INITIATION OF THE LAURENTIDE ICE SHEET

J.F. Royer, M. Déqué, P. Pestiaux

Centre National de la Recherche Météorologique
Av. Eisenhower Prolongée, 31057 TOULOUSE cédex France

Over long periods of time the Earth's climate is not stable but tends to oscillate between two different regimes corresponding to glacial and inter-glacial states. As we have been in the present interglacial regime since about 10 000 years, it seems a particularly interesting problem of climatic research to investigate the mecanisms responsible for the interglacial-glacial transitions.

The peak of the previous interglacial can be clearly identified in deep-sea cores (as isotopic stage 5e) and is dated at about 125 000 years B.P. Various palaeoclimatic indicators testify that the climate was then rather similar to the present climate (or even possibly warmer in some areas). The return to colder conditions with rapid accumulation of continental ice-sheets on the Northern Hemisphere took place somewhat before 115 000 B.P.

According to the astronomical theory of climate, of which the first quantitative formulation was given by M. Milankovitch, glacial oscillations have their origin in the insolation variations resulting from the orbital perturbations of the Earth. The summer insolation patterns at 125 ky and at 115 ky effectively show large deviations relatively to present values, due mainly to higher values of the eccentricity at this former period. The insolation is about 16 % smaller in July and larger in January at 115 ky than at 125 ky B.P., because the perihelion takes place in January instead of July.

In order to test the importance of such insolation variations for this par-ticular climatic transition, and better understand their direct influence on the atmosphéric circulation, we have performed by menas of a General Circulation Model (GCM) 2 simulations of a complete annual cycle each, with the insolation conditions computed respectively from the values of the orbital elements at 125 ky and 115 ky (according to Berger, 1978, J.A.S. 2362-2367). Surface boundary conditions where identical in the 2 experiments and specified from present-day climatology, with seasonal wariation of sea-surface temperature and ice-pack limits.

The model used for this experiment is the 10-level sigma-coordinate primitive equations, global spectral G.C.M. developped at Etablissement d'Etudes et de Recherches Météorologiques (E.E.R.M.). We used a version with truncation T10-13 and a computational grid of 20 x 32 points. Its physical parameter-izations include a simplified radiative scheme with diurnal variation of insolation, interactive cloudinee computed by the model from relative humidity values by an empirical formula, detailed hydrologic cycle and surface processes with evolution of soil temperature, water content and snow cover.

In the comparison of the statistics of the 2 experiments we found only very small changes in global-mean annual-mean parameters. However the seasonal cycle of several variables such as air-temperature and sea-level pressure (Fig. 1) averaged over the continents is significantly different. We can see on this figure that the summer in the Northern Hemisphere is much cooler (up to 6° C near 50° N) at 115 000 B.P., with higher surface pressure showing a reduction of the Monsoon phenomenon. Conditions in winter tend to be warmer only in the low latitudes in response to the higher insolation, so that in annual mean the Northern High Latitudes (53°-90° N) undergo a cooling of about 0.7° C.

We can notice the lower values of pressure in autumn and at the end of winter in the high latitudes that indicate more cyclonic conditions generally favourable to enhanced snow precipitations.

The geographical pattern of the variations from 125 ky to 115 ky shows in annual mean (Fig. 2) a region of surface cooling of more than 2° C over Canada, with an increase of soil water content reflecting a more positive blance of precipitation versus evaporation. Such result can be interpreted as a possible indication that the change in insolation from 125 ky to 115 ky hab by itself a significant effect on the atmospheric circulation leading to climatic conditions appropriate to the extension of permanent snow-cover over the Labrador area. According to the theories of instant glacierization, such persistence of the snowfields could be the initial step to trigger the formation of the Laurentide Ice-sheet.

In order to explain the subsequent growing of the ice-sheet other factors not considered in this experiment, such as the albedo feedback and modifications of oceanic circulation, will have to be taken into account. Other simulations with more realistic surface conditions for different phases of the ice-sheet growth, are necessary to analyse in more details the mecanisms responsible for the transition to glacial conditions. However, we consider that this simulation has shown clearly the essential role played by the insolation forcing and that this positive result should encourage further research on this subject.

A)

115-125 KA CONTINENT

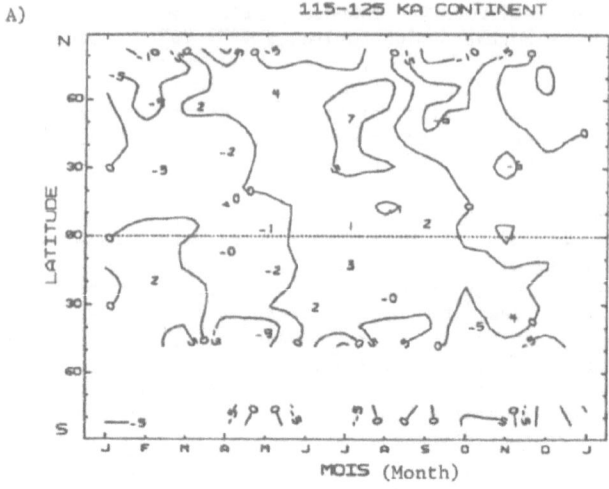

MOIS (Month)

· PMER -MB-

B)

115-125 KA CONTINENT

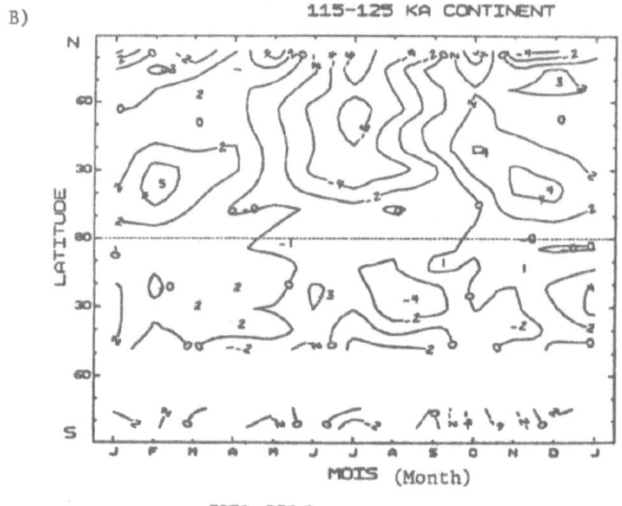

MOIS (Month)

T850 -DEG C-

Fig. 1 : Latitude-month distribution of the difference between
_____ the 2 simulations (115 ky minus 125 ky)
zonally averaged over the continents

A. (top) Sea-level pressure (mbar or 10^{-2} Pa)
B. (bottom) Temperature at 850 mb (°C)

A)

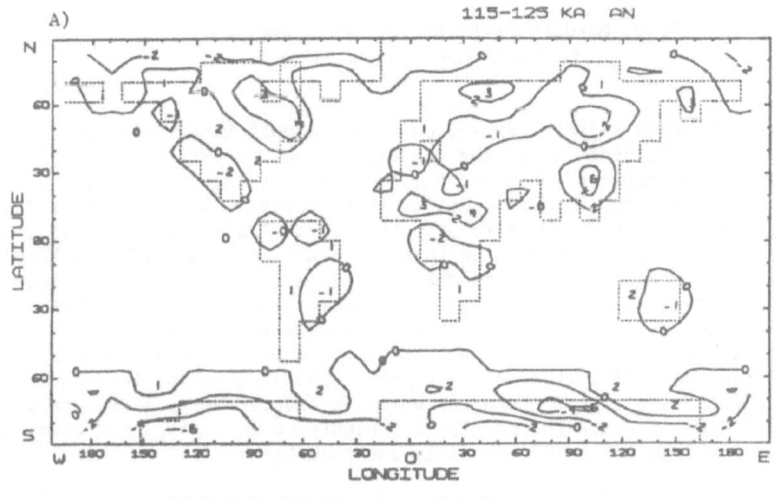

A)   TEMPERATURE DU SOL -DEG C-

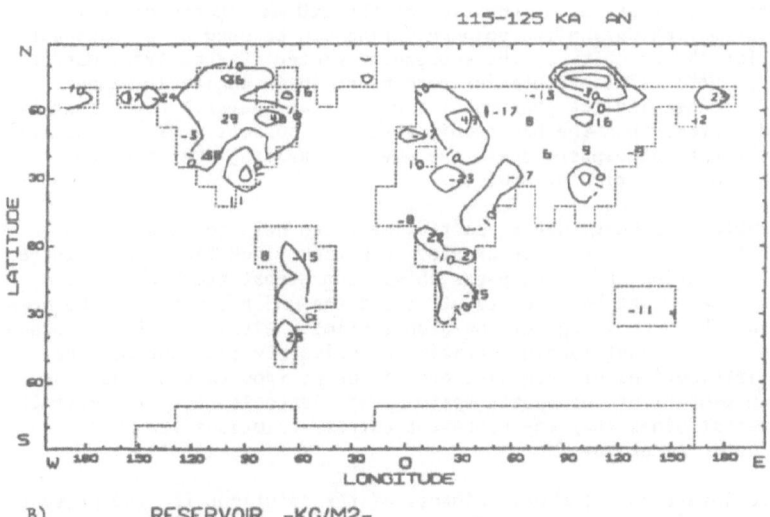

B)   RESERVOIR -KG/M2-

Fig. 2 :Geographical distribution of the difference of annual
———— means (115 ky minus 125 ky)

    A. (top) Surface temperature (°C)

    B. (bottom) Soil water content (kg/m^2)

PLANETARY WAVE CLIMATOLOGY EXPERIMENTS

By N. Murdoch, D.R. Davies, P.J. Everson
Exeter University, U.K.

In order to determine climatic evolution on a seasonal time scale, an understanding of the dynamics of the planetary waves is essential. At Exeter research is centered on using G.C.M.'s to investigate the factors influencing the phasing and amplitudes of these waves. There are two lines of approach in applying G.C.M.'s to climate study. The first is to use a highly detailed model in which the effect of anomalies (e.g. SST anomalies) as the net result of the complex feedback mechanisms within the model are evaluated. While this method is satisfying from the point of view that little is omitted from the model which emulates the real atmosphere to a close approximation, it has the drawbacks of being expensive in computer time and it can be difficult to perceive the mechanisms behind the resultant teleconnections. The second approach, which is adopted here, is to use a simplified G.C.M. in which the physics are highly parameterised and examine how various forcings act both separately and together.

Recently our interest has focussed on the investigation of atmospheric modes. A mode here is taken to mean a climatic regime in which the planetary waves (wave numbers $1 \rightarrow 4$) show persistence to systematic evolution over an extended period of time. The multiple equilibria solutions for barotropic flow demonstrate that the simplified atmospheric equations possess mode type solutions (Charney and Devoren 1979) and for the real atmosphere, there is some suggestion that mode type behaviour exists during the winter evolution; it has been found that the orientation of the 200 mb. October-November circum-polar vortex during the spin up period can be used as a reasonable indicator for the severity of the subsequent winter (Reeve, 1982, Murdoch and Davies, 1983). The orientation condenses, in an empirical manner, the information concerning the phasing of the planetary waves. Given that less simplified systems than the barotropic will also possess mode type solutions and that the real atmosphere does, the question arises, what are the determinants of a particular mode ?

It is possible that knowledge of past climate can help to answer this question. Different modes have existed on Milankovitch time scales where the solar insolation and also, presumably, mean global temperature were different. What might be more useful to aid the understanding of the present climate would be a knowledge of the global climate within the last thousand uears when the external forcings remain approximately constant but the internal distributions of anomalies are different from today. Thus the summers and winters of the Little Optimum and Little Ice Age, for example, are of interest since they may represent extreme solutions for the present set of climatic parameters.

In order to investigate the determinants of the solutions for the present climate parameters, several numerical experiments have been undertaken using a two-level quasi geostrophic, σ co-ordinate hemispheric spectral model. The aim of the experiments described below has been to examine how the orography and albedo morphology during the northern hemisphere winter influence the planetary waves. The variation in albedo around a latitude circle in the range 45° to 60° during the winter is quite large and this has a strong impact on the resultant heating of an atmospheric column with cooling occurring where the albedo is high.

The four experiments carried out are

(1) With no mountains and climatological albedo

(2) With mountains and a zonally averaged albedo

(3) With mountains and climatological albedo

(4) With mountains and interactively determined albedo distribution.

In each case the model was run for 130 days, with the same initial conditions. The first 50 days output was discarded and charts averaged over the remaining 80 days were produced. The heating function had winter time solar insolation data values and was responsive to the surface temperature via the surface balance condition. In run 4 the albedo was determined directly from the surface temperature; more sophisticated schemes which would include thermal inertia could be developed later. The climatological albedo used in run 1 and zonally averaged albedo distribution of run 2 are shown in Figure 1.

The 80 day average surface pressure and 200 mb. charts were produced. The 200 mb. chart was generated so that the circumpolar vortex orientation could be seen and compared with the real vortex whose orientation as mentioned above is an important feature of the winter climate.

Results

(1) The first run (no mountains, fixed climatological albedo) can be interpreted on the basis of there being a heat source where the albedo it least (Figure 2). High pressure is found over the reduced albedo with a low downstream. The 200 mb. chart shows a similar pattern with rigde and trough displacement to the west. This result agrees with the classic Smagorinsky (1953) solution for mid-latitude heat sources.

(2) The second run (mountains, zonally averaged albedo) shows the effects of the mountains with high pressure over both the Rockies and Himalaya (Fig. 3). The 200 mb. chart for this run reveals a far more asymmetric vortex than run 1. These are roughly equal east-west (which tends to be associated with a cold winter) and north-south components and no strong preferment is seen.

(3) The combined orographic and albedo induced forcings are present in run 3 (Fig. 4). The 200 mb. and surface charts are qualitatively more like those of run 2 than of run 1, indicating that it is the orography rather than the albedo which establishes the basic flow pattern. The surface pressure chart shows a high pressure system over Europe, which was not present with the symmetric albedo. This is interesting because during the present winter there has been an anticyclone over Europe which has displaced the het stream northwards bringing relatively warm air to the U.K. giving a mild winter; last winter (1981-1982) the high pressure system over Greenland dominated bringing cold air down on the U.K. as the jet was forced south. Thus it may be that the winter evolution is characterised by the relative strenghts of the Greenland and European highs. The results of run 3 show that the albedo distribution can influence these systems.

(4) In the fourth run the albedo was allowed to be interactive (Fig. 5). The
    200 mb. chart is interesting because it has far greater asymmetry than
    any of the previous runs. This suggests that perhaps while the albedo
    does not determine the possible atmospheric modes which are set yp by
    the mountains, it does pick out one which it then helps reinforce and so
    characterise the winter.

These experiments are preliminary and currently sensitivity studies are being
undertaken and the energetics are being investigated. It would be
interesting to run these experiments with parameters (albedo, surface
temperatures) from the Little Optimum and Little Ice Age and try to reproduce
the climate of those times.

Other experiments utilising an atmospheric G.C.M. to which an oceanic mixed
layer model is coupled are also being run. In these experiments the effects
of varying the lateral oceanic heat flux divergence in the modelled Pacific
and Atlantic oceans and the steady state solutions for the atmosphere and
ocean are being determined.

References

Charney, J.G. and Devore, J.G., 1979, "Multiple flow equilibria in the
            atmosphere and blocking", Journal of Atmospheric Sciences, 36,
            pp. 1205-1216.

Murdoch, N. and Davies, D.R., 1983, "Winter Prediction, 1982-83", Weather
            (Letter), Vol. 38, p. 33.

Reeve, C.E., 1982, "Winter 1981-82", Weather, (Letter), Vol. 37, p. 28.

Smagorinsky, J., 1953, "The dynamical influence of large scale heat sources
            and sinks on the quasi-stationary mean motions of the atmosphere",
            QJ. R. Met. Soc., 79, pp. 342-366.

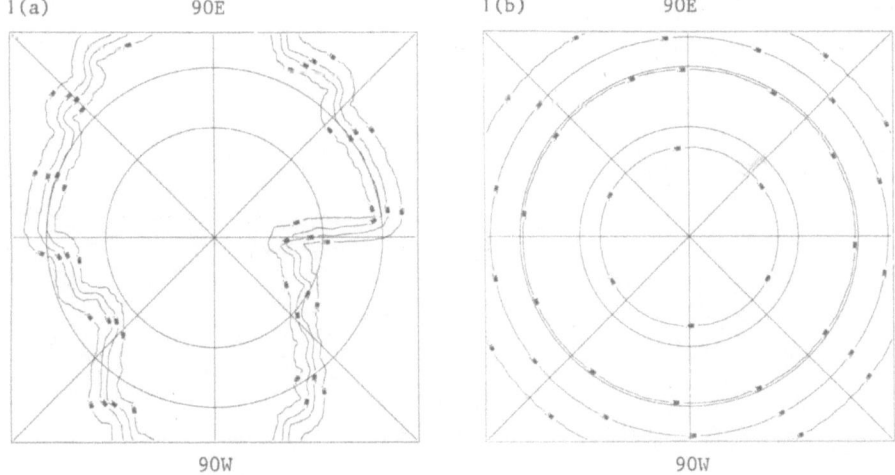

Figure 1 Stereographic projection of (a) the winter time albedo, and (b) zonally averaged albedo used in the experiments are shown.

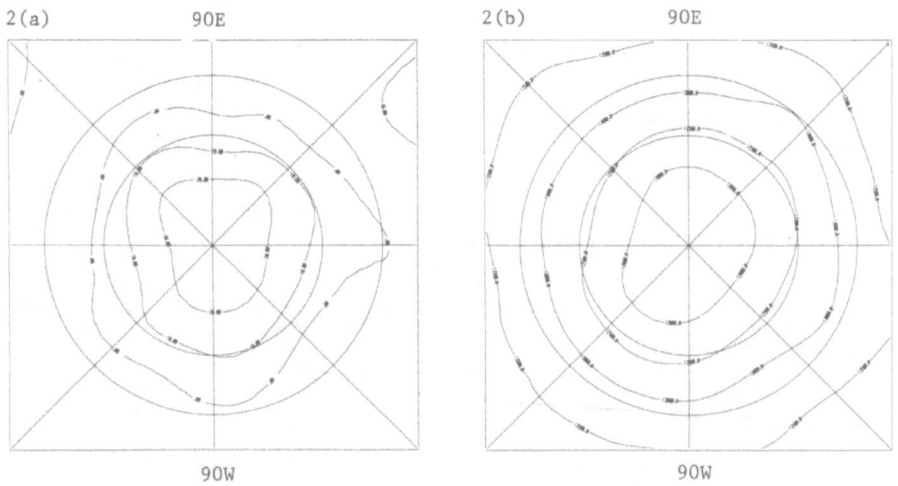

Figure 2 The surface pressure (a) and 200 mb. (b) charts from run 1 (no mountains, fixed climatological albedo (see Figure 1(a)) are shown. Latitude circles at 60° and 45° are drawn.

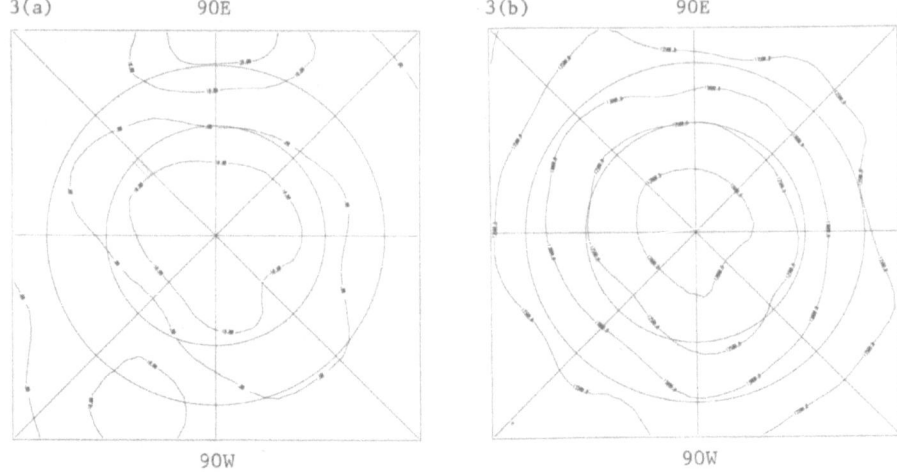

Figure 3 The surface pressure (a) and 200 mb. (b) charts from run 2 (mountains, zonally averaged albedo (see Figure 1(b)) are shown.

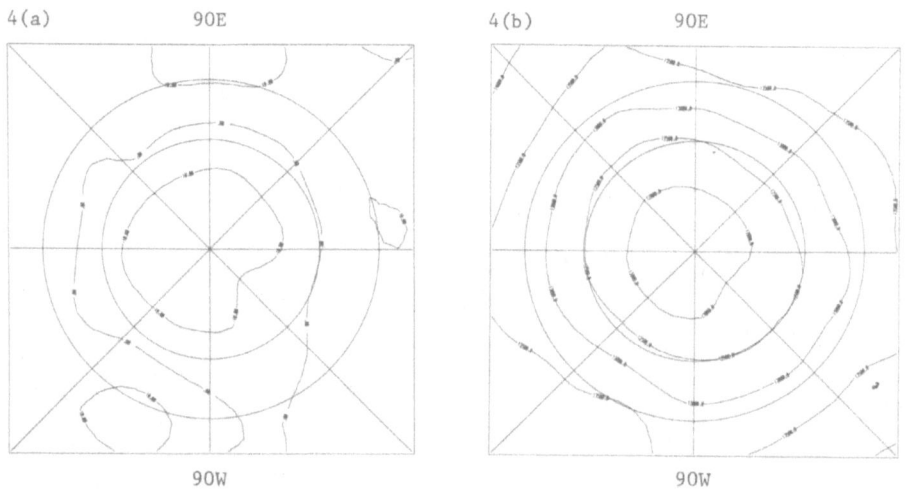

Figure 4 The surface pressure (a) and 200 mb. (b) charts from run 3 (mountains, climatological albedo) are shown.

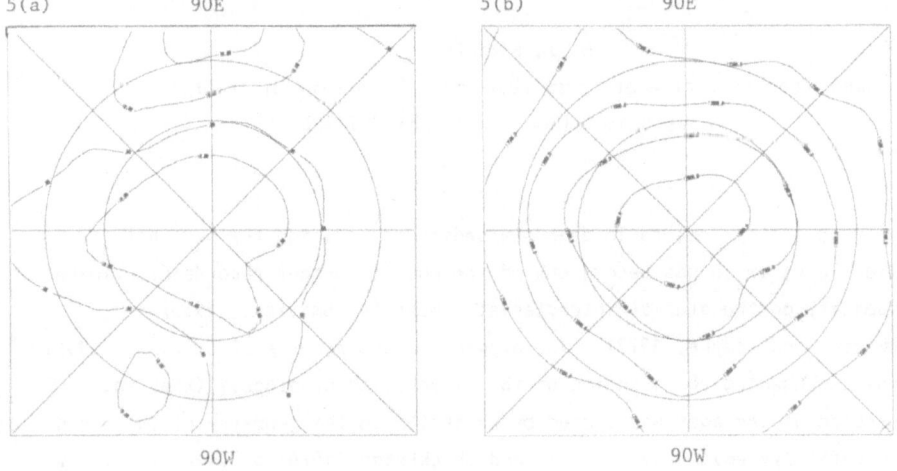

Figure 5   The surface pressure (a) and 200 mb. (b) charts from run 4
(mountains, interactive albedo) are shown.

THE EVOLUTION OF PLEISTOCENE CLIMATIC VARIABILITY

N. J. Shackleton

Sub-department of Quaternary Research, University of Cambridge.
Free School Lane, Cambridge CB2 3RS.

The oxygen isotope record in deep sea sediments covering the past half
million years or so has become one of the most important records of climate
variability on the glacial-interglacial timescale (Emiliani, 1955;
Shackleton and Opdyke, 1973). The oxygen isotope records of Emiliani (1955)
immensely enhanced the standing of the Milankovich hypothesis (that the
succession of ice ages was caused by variation in the geometry of the earth-
sun orbital system). Hays, Imbrie and Shackleton (1976) provided the first
more or less mathematically rigorous test of the hypothesis and again the
oxygen isotope record, an indicator of the Nothern hemisphere glacial record,
was vital to the success of their study which was based on Southern
hemisphere cores.

The study of early Pleistocene and pre-Pleistocene climate variability has
until recently been hampered by the lack of really suitable material for
study. Shackleton and Opdyke (1976) studied the lower Pleistocene in a very
long piston core (over 20 m) but their time-resolution was limited by a low
accumulation rate of only about 1 cm per thousand years; extending the
record to 3.5 million years (Shackleton and Opdyke, 1977) forced them to an
even lower accumulation rate and to further loss of resolution. Thus these
stratigraphically long records did not provide us with the detail that would
be needed to investigate the changing response of the global climate system
to astronomical forcing; this is a matter of some concern since we can never
be fully satisfied by the Milankovich hypothesis until we can satisfactorily
explain why it was only during the past million years or so that these
apparently subtle astronomical variations have given rise to the drama of the
ice ages.

Experience has shown that the best records for studying climatic variability
associated with astronomical forcing comes from areas of the deep sea floor
where sediment acculumates at around 4 cm per thousand years. At such an
accumulation rate, a three million year record covers 120 m and is quite
inaccessible by conventional piston coring. Rotary drilling in soft sediment

causes quite unacceptable disturbance, but recently a hydraulic piston corer has been developed which enables long, undisturbed sequences of soft deep-sea sediment to be recovered (see, for example, Prell and Gardner (1982)).

The first studies of the cores recovered by this new device are only now coming to fruition. Prell (1982) obtained a long isotope record from the Caribbean, the area from which Emiliani (1955) made his first classic studies. Shackleton and Hall (in press) obtained a very detailed record from a Pacific site (DSDP Site 504) and in this material unpublished studies by Shackleton in collaboration with N. G. Pisias (Oregon State University) have demonstrated very clearly that forcing of climatic variability by chages in the earth's orbital system was occurring well before the onset of Pleistocene glaciation. More recently an excellent site was cored using the HPC in the North Atlantic (DSDP Site 552A) and again Shackleton and Hall (in preparation) have obtained a detailed isotopic record in which mathematical analysis is under way in collaboration with N. G. Pisias and P. Pestiaux (Louvain-la-Neuve, Belgium).

Although the studies mentioned above are still unpublished it is evident that the study of high resolution HPC sections from the ocean basins will enable us to answer one of the crucial questions in the understanding of climate change. It is now generally agreed that changes in the earth's orbital geometry were the primary cause of the succession of glacial-interglacial changes during the past million years, and a great deal of good work is now going into attempts to use this knowledge to improve our understanding of climate. Any model that is used to predict the effect during the next century of man's activities (for example, adding carbon dioxide to the atmosphere) will be regarded as more credible if it can successfully predict the effect of subtle changes in the earth's orbital geometry. However, climate models are notoriously sensitive to some of the parameters specified. The real climate system, like the numerical model, must be very sensitive to certain parameters. By studying the geological record, we must learn which physical parameters changed during the past three million years and why these physical changes had such a dramatic effect on the sensitivity of the real climate system to orbital variations.

Emiliani, C. (1955). Pleistocene temperatures. Journal of Geology 63, 538-578.

Hays, J.D., Imbrie, J. and Shacketon, N.J. (1976). Variations in the earth's orbit : pacemaker of the ice ages. Science 194, 1121-1132.

Prell, W.L. (1982). Oxygen and Carbon Isotope Stratigraphy for the Quaternary of Hole 502B : Evidence for two modes of isotopic variability. Initial Reports of the Deep Sea Drilling Project, vol. 68, 455-464.

Prell, W.L., Gardner, J.V. et al. (1982). Initial Reports of the Deep Sea Drilling Project, vol. 68 : Washington (U.S. Govt. Printing Office).

Shackleton, N.J. and Hall, M.A. (1983). Stable Isotope Record of Hole 504 Sediments : High Resolution Record of the Pleistocene in Cann, J.R., Langseth, M.G. et al., Initial Reports of the Deep Sea Drilling Project, vol. 69. Washington, U.S. Govt. Printing Office.

Shackleton, N.J. and Opdyke, N.D. (1973). Oxygen isotope and palaeomagnetic stratigraphy of Equatorial Pacific core V28-238 : oxygen isotope temperatures and ice volumes on a 10^5 year and 10^6 year scale. Quaternary Research 3, 39-55.

Shackleton, N.J. and Opdyke, N.D. (1976). Oxygen isotope and palaeomagnetic stratigraphy of Equatorial Pacific core V28-239, Late Pliocene to Latest Pleistocene. In Investigation of Late Quaternary Paleoceanography and Paleoclimatology, ed. R.M. Cline and J.D. Hays, 449-464. Geological Society of America Memoir 145.

Shackleton, N.J. and Opdyke, N.D. (1977). Oxygen isotope and palaeomagnetic evidence for early Northern Hemisphere glaciation. Nature 270, 216-219.

Session C : Glaciated polar regions and their impact
on global climate

HISTORY OF THE NORTH POLAR SEAS DURING THE PAST 5 MILLION YEARS

Jörn Thiede, Department of Geology and Paleontology, Kiel University,
Olshausenstrasse 40/60, D-2300 Kiel, F.R. Germany

For many years we did not have any opportunity to investigate the late
Cenozoic palaeoclimatic history of the Arctic Ocean per se because it was
impossible to collect suitable sample material. Early sampling efforts
have been carried out by F. Nansen, but long sediment cores which allow
palaeoclimatic and palaeo-oceanographic studies have only been collected in
the sixties from the ice island T3 (Hunkins et al., 1971). In the meantime
these sample collections have been supplemented by Canandian, U.S., Norwegian,
Swedish and Soviet expeditions and numerous sediment cores are available for
stratigraphic studies (Clark et al., 1980, Boshöm 8 Thiede, 1981). However,
it has been shown that the present cores comprise (Fig. 1) only of upper
Cenozoic deposits (Pleio-Pleistocene, possibly a bit of uppermost Miocene)
except two short core segments with older tuffaceous and siliceous muds which
are found displaced into Pleisticene sediments. The two older core segments
are of Palaeocene and late Cretaceous age, and both of them indicate a
temperate climatic regime. As the oldest Neogene sediments from the Arctic
Ocean already contain ice-rafted coarse terrigenous debris the important
processes during the initiation and early stages of Northern Hemisphere
cooling and glaciation cannot be reconstructed at all.

To resolve the younger palaeoclimatic history of the Arctic based on Arctic
Ocean sediment cores has also encountered considerable problems because all
presently available dated sediment cores have given very low sedimentation
rates and heceforth no detailed stratigraphic resolution. The fossil record
of these sediments is intermittend but at times very high, though the
dominant components of the fossil faunas are essentially endemic to polar
water masses.

The Neogene sediments of the Arctic Ocean in general consist apparently of
very finegrained reddish-brownish to greyish rather homogenous terrigenous
muds. Detailed lithostratigraphic studies, both in the eastern and western
Arctic basins have shown that a number of characteristic lithostratigraphic
units can be defined bases on their grain size, fossil contents, colour,
which can be correlated over wide distances, across several basins and which
therefore seem to indicate that basinwide changes of the depositional
environment in the Arctic Ocean happened several times during the Neogene.
However, differences of the nature of the surface sediment layer of the
western and the eastern Arctic Ocean suggest that in this way defined stratigraphic boundaries do not represent isichronous interfaces and changes of
the depositional environment, but taht they rather illustrate their trans-
gresive nature. New comparative stratigraphic studies of sediment cores
from the eastern and western Arctic basins will have to elucidate this
problem.

The upper Miocene to Recent sediments which have successfully been dated up
to now (Clark et al., 1980) comprise of 6-7 layers of enriched in coarse
ice-rafted terrigenous debris and somewhat fewer fossil-rich horizons.
Based on magnetic data it is clear that the events which led to the changes
of sediment composition cannot easily be reconciled with a glacial/inter-
glacial stratigraphy. However, both observations pose very interesting
palaeoclimatic and palaeooceanographic problems.

The surface sediments in the western Arctic basins seem to contain numerous fossils (Herman, 1974) whereas the youngest deposits of the eastern Arctic basins seem to be devoid of fossils (Markussen et al., 1982). Otherwise, the subsurface sediments in the entire Arctic Ocean consist of alternating layers of fossil-rich and -poor (or even barren) deposits. In the entire basin we observe simultaneous increases and decreases of planktonic (mainly foraminifers) and benthic (mainly foraminifers, ostracods, molluscs, echinoderms) remains; henceforth, it can be assumed that the entire marine food web responded to specific oceanographic/climatic settings which allowed long spells of relatively high productivity in the Arctic Ocean. It is intriguing to speculate over the oceanography of these time spans and over the observation that only negligable quantitues of fossils produced by primary producers have been found, whereas almost the entire fossil record of Arctic deep-sea sediments consists of fossils produced by secondary producers or organisms of higher levels of the marine food web. Productivity patterns of the modern Arctic Ocean are not sufficiently well known to be able to judge if the fossil record corresponds to the original signal of productivity or if the dominance of animal remains is an artifact of differential pre-servation.

The other observation of particular interest is the occurrence of layers enriched in ice-rafted coarse terrigenous sediment components (Clark et al., 1980; Thiede et Markussen, 1983) which have been encountered, even over structural highs in the central Arctic Ocean far away from any land region. These ice-rafted components are .5-80 mm in diameter (maybe much larger), their shape can be rounded or not; they may have a very variable composition of crystalline or sedimentary rocks. They are found embedded into a fine-grained clayey matrix in horizons up to several decimetesr thick. If the sedimentation rates of the cores presently available are as low as suggested by their magnetic stratigraphy, then these parts of high inputs of ice-rafted material must have lasted for several hundred thousand years at a time, and they must have been generated by an oceanographic setting which has recurred 6-7 times during the past 5 million years. The source regions of this material are believed to be located on the continents sur-rounding the Arctic Ocean, but as long as we are unable to define their transport mode (by coastal ice, riverine ice, glacier ice or ice from ice-shelves) and the climatic/oceanographic scenario resulting in melting the transporting ice, we will not be able to understand the late Cenozoic palaeo-oceanography of the Arctic Ocean.

References

Boström, K. and J. Thiede, 1981, Marin geologi och geofysik i Arktiska oceanen - problem och preliminära resultat. "Ymer" 101, pp. 90-109.

Clark, D.L., R.R. Whitman, K.A. Morgan and S.D. Macley, 1980, Stratigraphy and Glacial Marine Sediments of the Amerasian Basin, Central Arctic Ocean. Geol. Soc. Amer. Spec. Pap. 181, 57 p.

Herman, Y., 1974, Arctic ocean Sediments, Microfauna, and the Climatic Record in Late Cenozoic Time, pp. 283-348 in Y. Herman (ed.), Marine Geology and Oceanography of the Arctic Seas, 397 p., (Springer Verlag), New York.

Hunkins, K., A.W.H. Bé, N.D. Opdyke and G. Mathieu, 1971, The Late Cenozoic History of the Arctic Ocean, pp. 215-237, in K.K. Turekian, The Late Cenozoic Glacial ages, 606 p., (Yale Univ. Press), New Haven.

Thiede, J. and B. Markussen, 1983, Istransportert droppsten over Polhavet, "Ymer", in press.

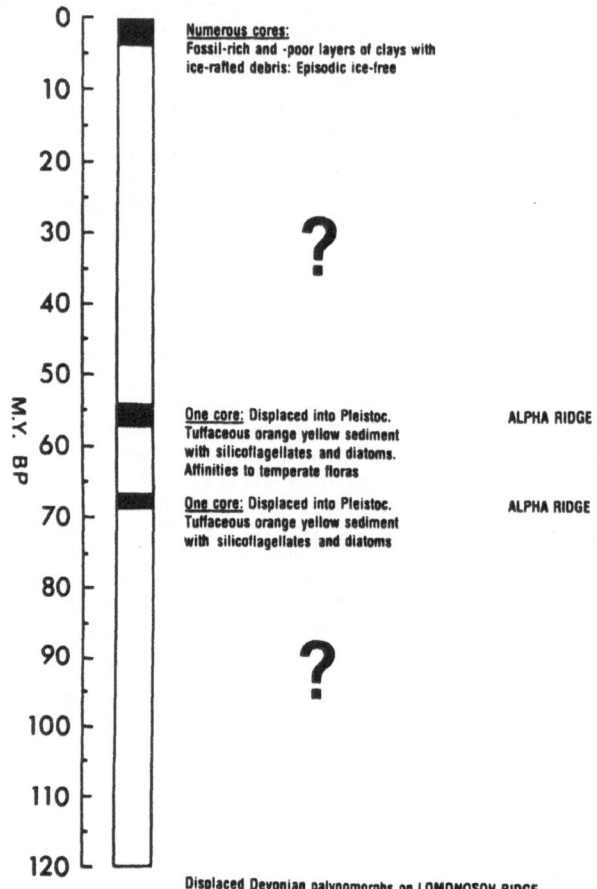

SEDIMENTS FROM THE ARCTIC OCEAN

Numerous cores:
Fossil-rich and -poor layers of clays with
ice-rafted debris: Episodic ice-free

?

One core: Displaced into Pleistoc.   ALPHA RIDGE
Tuffaceous orange yellow sediment
with silicoflagellates and diatoms.
Affinities to temperate floras

One core: Displaced into Pleistoc.   ALPHA RIDGE
Tuffaceous orange yellow sediment
with silicoflagellates and diatoms

?

M.Y. BP

Displaced Devonian palynomorphs on LOMONOSOV RIDGE

Fig. 1. Temporal distribution of presently available sediment
cores from the Arctic Ocean. Compiled from various sources.

SENSITIVITY OF GENERAL CIRCULATION MODELS
TO CHANGES IN SEA-ICE COVER

by T.S. Hills (UK Meteorological Office)

The last ten years have seen a number of experiments with atmospheric general circulation models (GCMs) to investigate the effect on the modelled atmosphere of changing the prescribed sea-ice cover. Tes responses obtained were often zonally asymmetric, demonstrating the limitations implicit in conducting zonal mean studies. For example, Newson (1973) found that by completely removing ice cover in the northern hemisphere in winter, in addition to the expected great warming at low levels in the Arctic, there was about a 1K cooling in the zonal mean at 30-50° N (Fig. 1). However, this was a manifestation of much larger cooling locally over the United States (8K), E Siberia (6K) and N Europe (2K) (Fig. 2). It was found that the Atlantic low moved SW when the ice was removed, reducing the warm low level westerlies over W Europe. Warshaw and Rapp (1972) however, in similar experiments with a 2-level model, found a zonal mean warming extending further south (Fig. 3), which they attributed to the outbreaks of polar air being less severe. It was not stated how much of this warming could be because of local heating, resulting from the removal of ice in Hudson Bay, for example. Flechter et al. (1973), using the same model as Warshaw and Rapp, found a smaller meridional temperature gradient tending to reduce baroclinic instability and the zonal mean westerles, but the low level heating and slight high level cooling greatly reduced static stability which has the effect of increasing baroclinic instability. These effects are commonly seen in such experiments. Also frequently seen is a shifting of the Atlantic depression tracks. Depressions tend to stay further south, tracking east to the Mediterranean, or to move north-west to the west of Greenland (e.g. Flechter et al., 1973 and Newson, 1973).

Other experiments have been run with less drastic changes than removing sea-ice completely. Herman and Johnson (1979) ran two experiments, one with the 16 year maximum January-February ice-cover at each grid point, the other with the minimum at each point recorded during 16 January-February periods (Fig. 4). Most of the effects of reduced ice-cover are consistent with those obtained by removing all the ice, except that they found that the depression tracks were further north. Also there were marked low latitude changes in mid-tropospheric temperatures; 500 mb temperatures increased by over 3K off W Africa and decreased by more than 3K off the west coast of N. America and over northern China (Fig. 5).

Williams et al. (1974) performed experiments with January and July conditions for present day ice-cover and for the ice and other surface conditions estimated to be present at the glacial maximum 20 000 years B.P. In January under ice-age conditions the Aleutian and Icelandic lows were 10° further south than in present-day conditions, with the depression tracks further south. There was a little less precipitation, especially in 0-10° N and 55-70° N. In July much less precipitation fell north of 10° N in the ice-age experiment with high pressure established over Asia. Also the zonal wind was more comparable with winter present day values. There was little cyclonic activity near the Laurentide ice sheet, but a depression track ran from eastern Europe into Asia.

Simmonds (1979) describes an experiment using a model of the southern hemisphere run using mean September ice conditions. The integration was repeated using March ice conditions instead. The upper westerlies were weakened over the anomaly area, and strengthened over Antarctica. The sea-level pressure patterns in the anomaly run showed increases in middle and high latitudes, rather than the decreases seen near the anomaly in other (northern hemisphere) experiments. The change was reminiscent of the observed pressure changes from September to March. The total latent heat flux from the southern hemisphere surface was found to be largely independent of the sea ice cover.

This is confirmed by an experiment at the UK Meteorological Office where July to September of one annual cycle integration with a 5-level GCM was repeated with Antarctic sea ice removed north of 60° S (Fig. 6). Figure 7 shows that sensible heat flux out of the surface was very large where the ice was removed, but the flux decreased over broad regions further north particularly where the low level flow had changed so as to bring more air from the north, and where air from the south had already been warmed by fluxes from the ice-free region (the largest upward fluxes in both integrations were just off the ice-edge). Contrary to Simmonds' findings, sea-level pressure decreased over the anomaly area (Fig. 8), but it must be borne in mind that Simmonds' experiment was conducted for a period nearer the equinox. In the Meteorological Office experiments the southern mid-latitude westerlies were weakened by up to 3 ms^{-1} ($\simeq$ 30%) at 700-500 mb and the low-level easterlies near the Antarctic coast were up to 3 ms^{-1} ($\simeq$ 30%) weaker (Fig. 9). Warmings were large (8-12K) at the lowest level over the anomaly area, and in excess of 4K even over the Antarctic continent (Fig. 10). Higher up there was a general cooling of up to 2K in the middle troposphere.

References

Fletcher, J.O., Mintz, Y., Arakawa, A. and Fox, T., 1973, Numerical Similation of the Influence of Arctic Sea Ice on Climate. Proceedings of the IAMAP/IAPSO/SCAR/WMO Symposium "Energy Fluxes over Polar Surfaces", Moscow, 1971.

Herman, G.F. and Johnson, W.T., 1977, The Effect of Extreme Sea-Ice Variations on the Climatology of the Goddard General Circulation Model. Sea-ice processes and models, ed. Pritchard, R.S. University of Washington Press.

Newson, R.L., 1973, Response of a General Circulation Model of the Atmosphere to Removal of the Arctic Ice-cap. Nature, 241, pp. 39-40.

Simmonds, I., 1979, The effect of sea ice on a general circulation model of the Southern Hemisphere. Symposium Proceedings "Sea Level, Ice and Climate Change", IAHS Pub. (nr. 13).

Warshaw, M. and Rapp, R.P., 1972, An Experiment on the Sensitivity of a Global Circulation Model. J. Appl. Met., 12, pp. 43-49.

Williams, J., Barry, R.G. and Washington, W.M., 1974, Simulation of the Atmospheric Circulation Using the NCAR Global Circulation Model with Ice Age Boundary Conditions. J. Appl. Met., 13, pp. 305-317.

FIGURE 1

Zonal mean temperature changes from the Newson (1973) experiment.

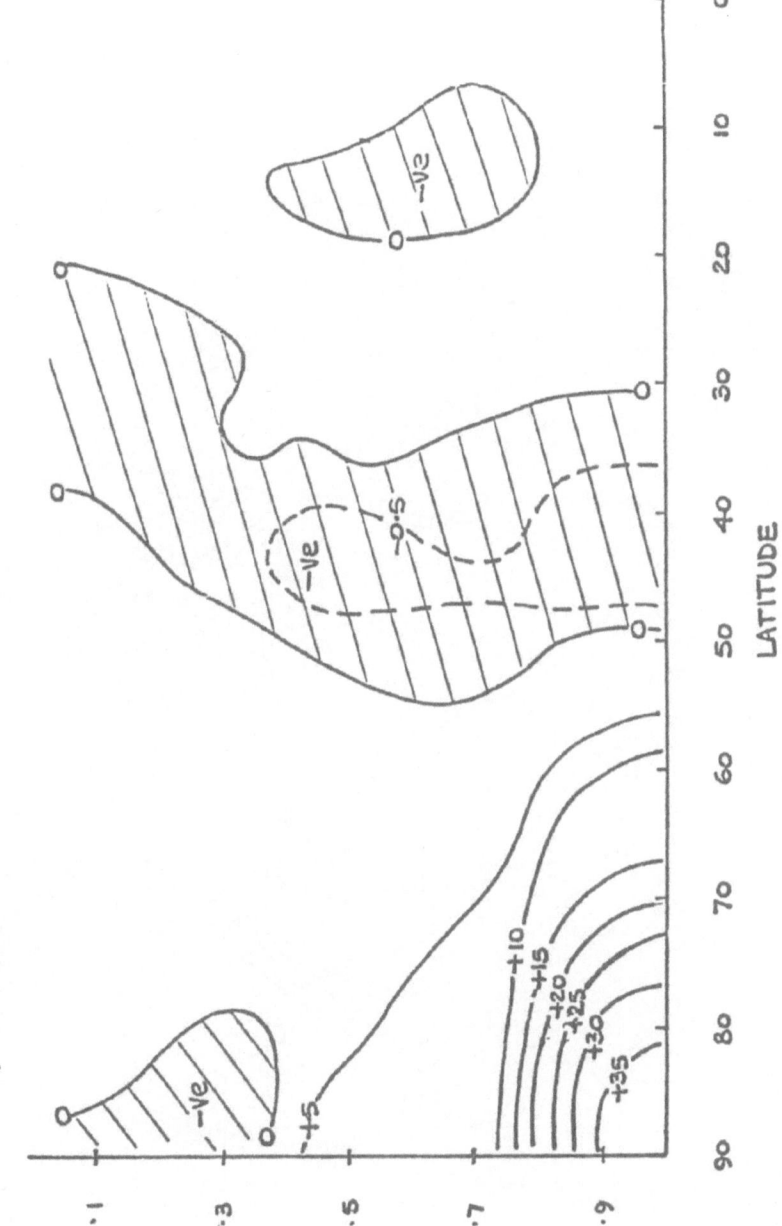

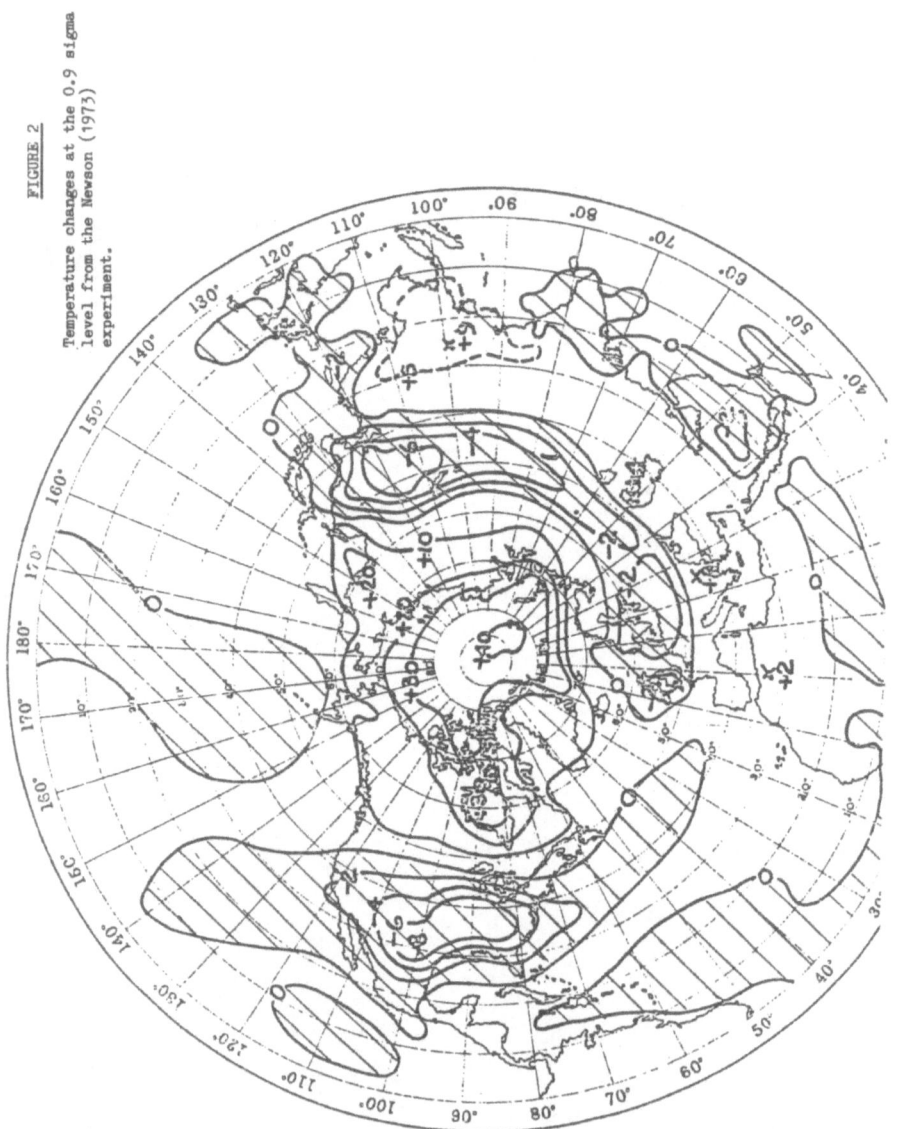

FIGURE 2

Temperature changes at the 0.9 sigma
level from the Newson (1973)
experiment.

FIGURE 3   Zonal Mean changes from Warshaw and Rapp (1972).

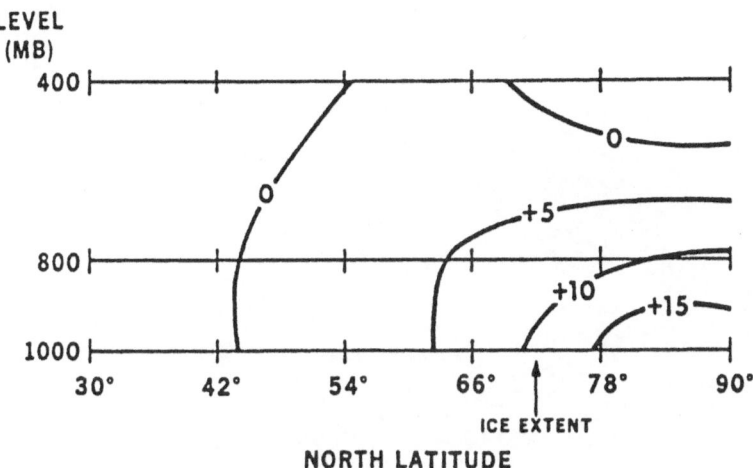

-- Temperature differences
(°C); (ice out) - (ice in).

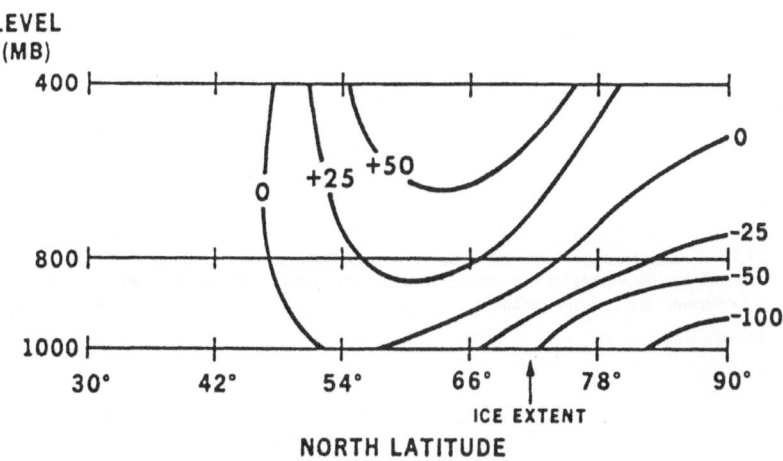

-- Geopotential-height differences
(m); (ice out) - (ice in).

REGIONS OF ICE MARGIN VARIATION (EXTREME)

Fig. 4. Location of ice margin variations. Regions enclosed by solid lines are areas that become covered with sea ice to construct "maximum" (i.e., extreme) ice conditions from "minimum" (i.e., control) conditions.

500 MB TEMPERATURE DIFFERENCE (°C)

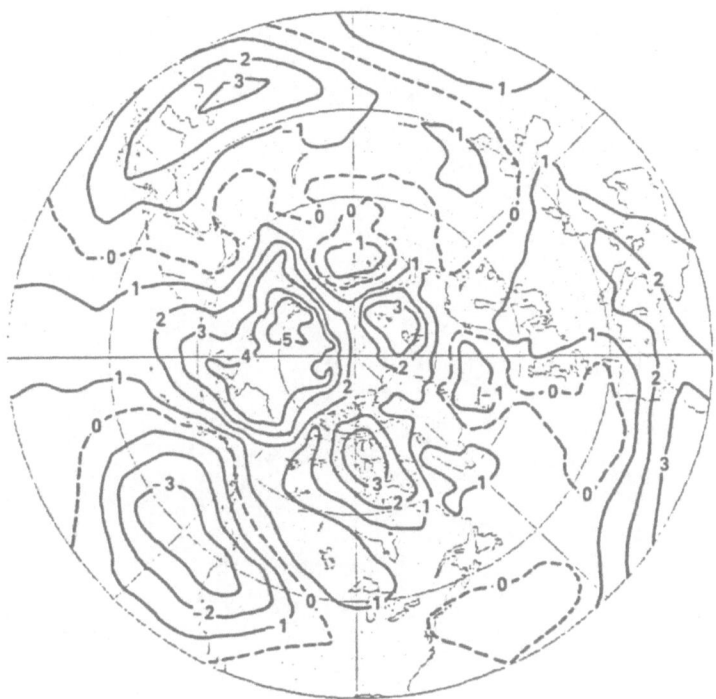

FIGURE 5

Difference (minimum minus maximum conditions) of 500 mb mean monthly temperatures.

Figure 6

Sea, land and sea-ice distribution in the southern hemisphere in
the control run in August. The annular region outlined is where
sea-ice was replaced by sea at 271.2K in the anomaly run.

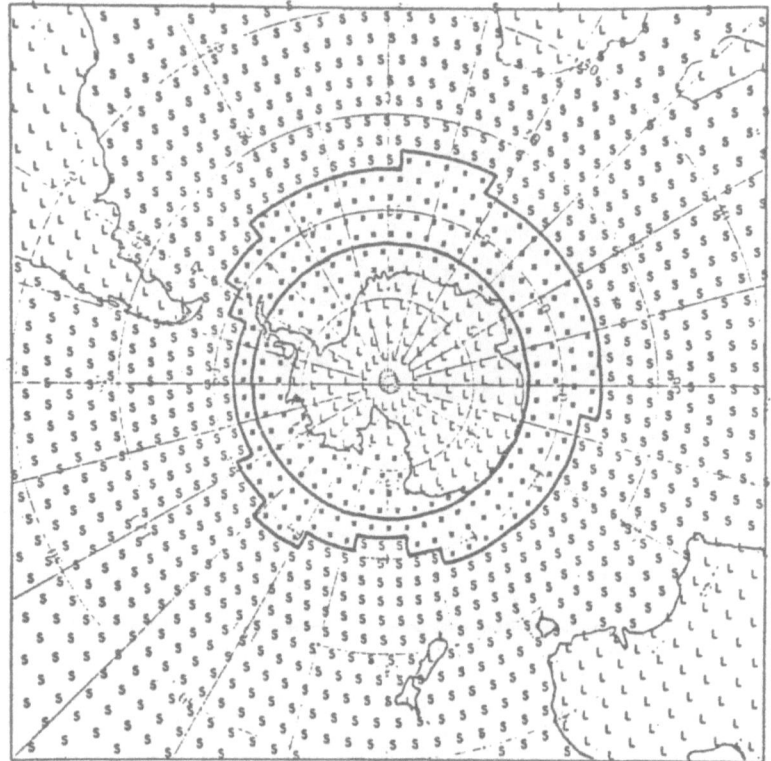

S.HEM. SURFACE CLASSIFICATION CHART
SEA ICE=*. SEA=S.LAND=L

EXPT 841DAY 250

Figure 7

DIFFERENCE IN SENSIBLE HEAT $(Jcm^{-2}day^{-1})$
BETWEEN SEA-ICE ANOMALY EXPERIMENT AND CONTROL   -   AUGUST

Negative areas are shaded

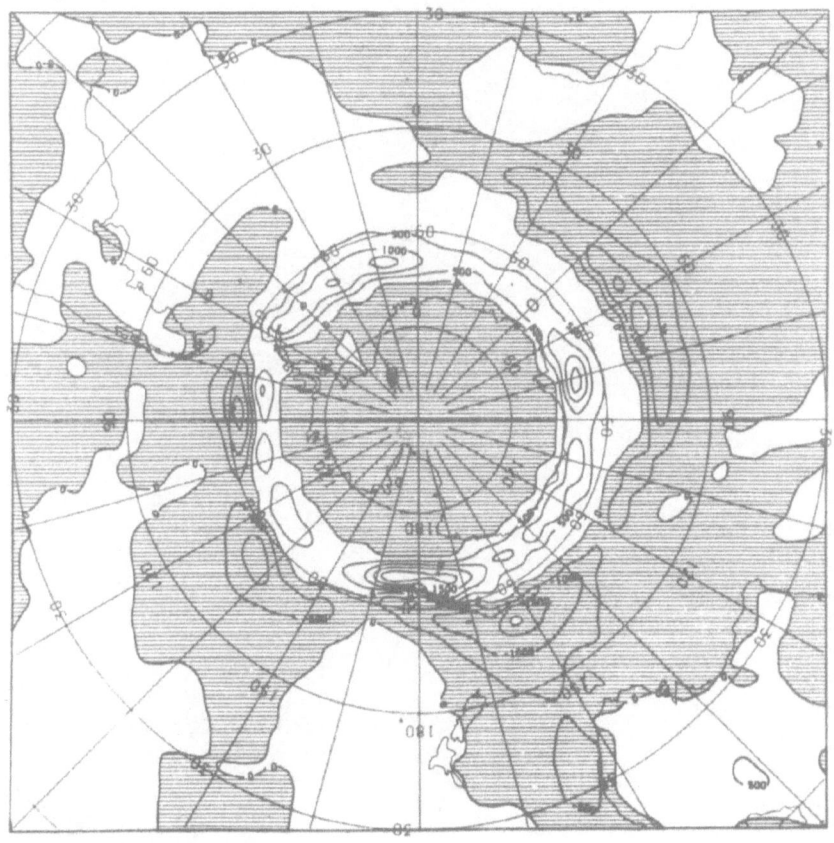

Figure 8

MEAN SEA LEVEL PRESSURE DIFFERENCES (mb)
BETWEEN SEA-ICE ANOMALY EXPERIMENT AND CONTROL - AUGUST

Negative areas are shaded.

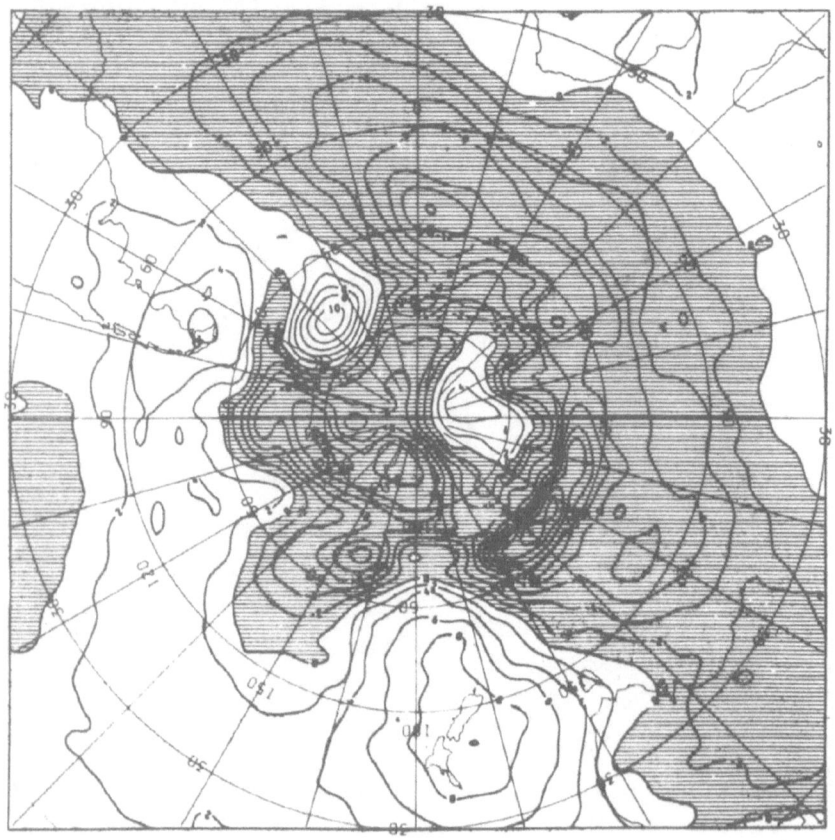

FIGURE 9

CHANGES IN ZONAL WIND (METRES PER SEC)
CONTOUR INTERVAL = 0.5 M/S^{-1}
AREAS OF DECREASE ARE SHADED
ZONAL AVERAGE

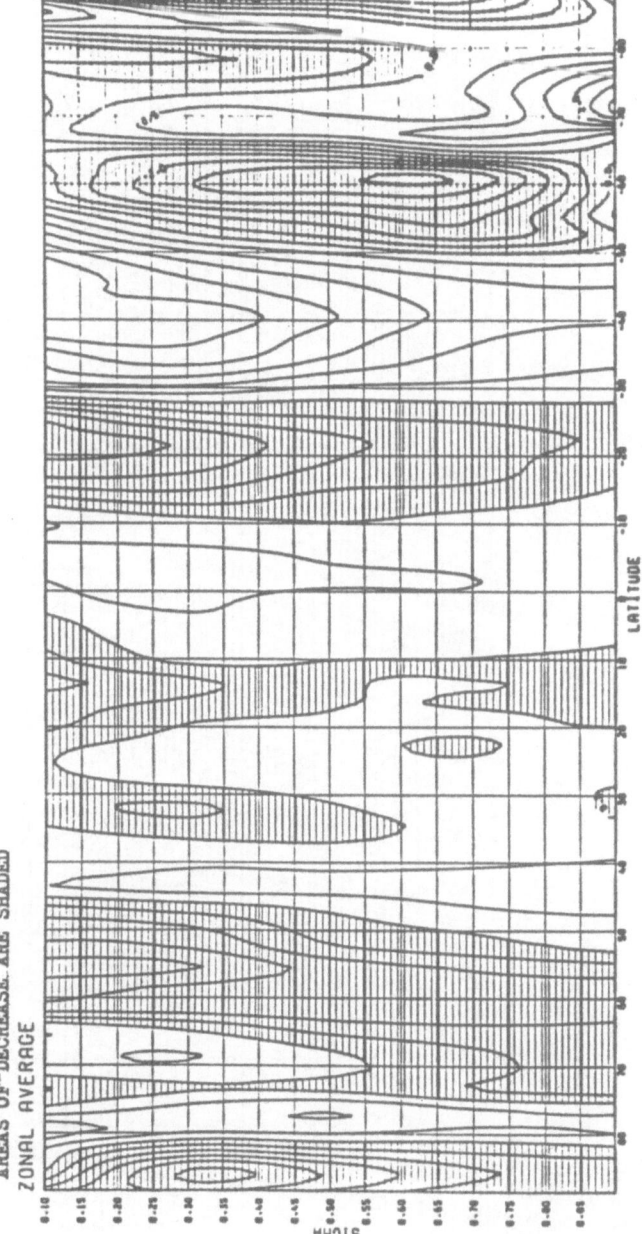

FIGURE 10

CHANGES IN TEMPERATURE(KELVIN)

CONTOUR INTERVAL = 2K
AREAS OF DECREASE ARE SHADED
ZONAL AVERAGE

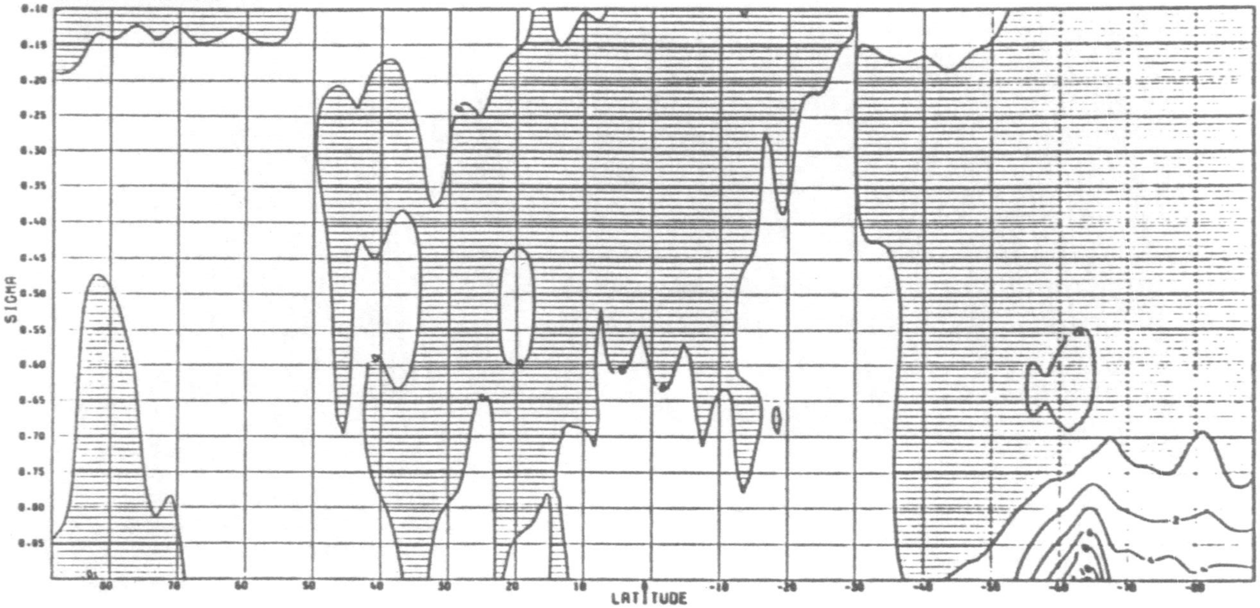

NUMERICAL MODELLING OF ARCTIC SEA ICE : REVIEW AND PRELIMINARY RESULTS

J.P. van YPERSELE

Institut d'Astronomie et de Géophysique
Université Catholique de Louvain
Chemin du Cyclotron 2
B-1348 Louvain-la-Neuve
Belgium

1. INTRODUCTORY REMARKS ON SEA ICE AND CLIMATE

According to Untersteiner (1975), the total amount of water in all earthly forms is estimated to be 138 x 10 km . Of this, 97.4 % is sea water; 0.0009 % is atmospheric water vapour; 0.5 % is ground water, mostly at great depths; 0.1 % is contained in rivers and lakes, and 2.0 % is frozen. Today, perennial ice covers 11 % of the erath's land surface and an average of 7 % of the world ocean.

Table 1, adapted from Untersteiner (1975), summarizes the distribution of the land and sea ice.

	Continental ice	Sea ice			
		Southern Ocean		Arctic Ocean	
		min.	max.	min.	max.
V (Volume in km^3)	30.94×10^6	5 000	30 000	20 000	50 000
A (Area in km^2)	16.2×10^6	2.5×10^6	20.0×10^6	8.4×10^6	15×10^{-6}
H (average thickness, $\frac{V}{A}$, in m)	1910	$\times 10$ 2.0	1.5	$\times 2$ 2.4	3.3
R (average thickness to diameter ratio) $(R = \frac{1}{2} H^{1/2} A^{-1/2})$	400×10^{-6}	1.1×10^{-6}	0.3×10^6	0.7×10^{-6}	0.8×10^6

The large differences observed in the average thickness to diameter ratio of land ice versus sea ice explains the high climatic sensitivity of sea ice extent as compared to land ice sensitivity : a one km^2 increase in sea ice area needs a much smaller energy deficit than a similar increase in continental ice extent.

Sea ice in turn, affects the climate system mostly in two ways :

- through the high albedo of ice cover, which significantly diminisches the surface absorption of solar radiation

- through the isolating influence of sea ice on the heat and momentum exchanges of ocean with atmosphere.

An interesting argument was developed by Budyko (1974, p. 265) to illustrate the importance of sea ice in the climatic system. The "Budylo's paradox" goes like this : at the Equator as well as in the Central Arctic, changes in the heat content of the earth-atmosphere system in June are small in comparison with the income of solar radiation. This is because this component of the heat balance reaches appreciable values only with substantial changes in ocean temperatures during the annual cycle. But this temperature changes little, either near the Equator (because of the lack of significant annual variation in the mean altitude of the Sun) or in the central region of the Arctic Ocean (because of the isolating influence of sea ice on the heat exchanges between ocean and atmosphere).

If the albedo of the earth-atmospher-system was the same at the Equator and in high latitude, the air temperature near the surface in June at high latitudes should be higher than the temperature near the Equator in order to compensate, through a higher radiation emission to space, for the horizontal flux of heat which is on the average directed from Equator to pole. Such a temperature gradient being the opposite of what is observed in reality, Budyko deduces that the explanation of this paradox lies in a significant difference in the albedo of the earth-atmosphere system with and without ice cover. Indeed, the mean value of albedo at 80° N turns out to be about 0.6, while the mean albedo for regions free of ice equals only 0.30.

2. MAIN SEA ICE PROCESSES

These processes can be subdivided into thermal and mechanical ones :

2.1. Thermal processes

- Sea ice has not real "melting point", because the size of the brine inclusions is a function of temperature; when this brine volume becomes too large (around -1° C to -0.1° C for the upper surface ice, depending on salinity), the ice disintegrates (Untersteiner, 1975).

- Two competing processes act to produce the "equilibrium thickness" of the sea ice : bottom accretion of ice in winter and (top and bottom) ablation in summer. Lateral melting can also take place for isolated floes.

- Sea ice conducts heat, and its thermal conductivity depends on temperature (due to the brine pockets). Transport by conduction is mainly done upward, when the upper surface temperature becomes sufficiently inferior to the bottom surface temperature (which is fixed at -1.8° C, the freezing point of normal seawater).

- Energy is flowing upward from the ocean mixed layer to the ice. This oceanic heat flux is estimated to be 2 W m^{-2} in the Arctic Ocean.

- Starting from an ice free ocean, ice forms at the surface when the entire mixed layer column is cooled to its freezing point. This is due to the fact that water with a salinity higher than 25 ‰ has a maximum in density below the freezing point; therefore water reaching the freezing point sinks and is replaced by warmer water until the entire column is cooled.

2.2. Mechanical processes

Sea ice forms a complex three-dimensional continuum that moves in response to wind and water stresses. It is generally agreed that the dynamics of sea ice on a large scale can be characterized using the following elements (Hibler, 1980) :

- a momentum balance describing ice drift, including air and water stresses, Coriolos force, internal ice stress, inertial forces, and ocean currents effets. A "rule of thumb" is that, with respect to the dominant components of this balance, much of the ice drift far from shore takes place nearly parallel to the isobars with a speed of about one-hundredth of the geostrophic wind speed. Ice stress terms tend to increase that parallelism during very compact ice conditions; when large numbers of leads and polynyas exist (cfr next paragraph), a drift angle op up to 30° right of the wind has been observed in the Arctic.

- an ice rheology which relates (on some suitable continuum space and time scale) the ice stress to the ice deformation and strength. The main problem in devising a large-scale constitutive law for pack ice is that tests on laboratory-sized samples cannot be directly applied to the mechanisms controlling floe-to-floe interaction on the large scale (Untersteiner, 1975). Significant mechanical events render the task difficult : pack ice cracks into plates of various sizes, which superpose on each other (ice rafting) or collide and form pressure ridges which are up to 25 meters high (ice ridging). Sea ice is also broken up by leads (fractures through sea ice widened by wind and water stress) and open areas ("polynias").

- an ice strenght determined primarily as a function of the ice thickness distribution, is finally needed as an input to the constitutive law.

3. MODELLING SEA ICE : A FEW EXAMPLES

Attempts to model the large scale sea ice behaviour reflect both points of view : the thermodynamical one and the dynamical.

The thermodynamical sea ice models are illustrated by the Maykut and Untersteiner (1971) one-dimensional model, which computes the time-dependent ice-thickness and vertical ice temperature profile of level ice forced with energy fluxes taken from the climatology. It is the most complete one-dimensional ice model devised so far, but the long time of integration it requires precludes its use in three dimensional climatic simulations. Semtner (1976) has significantly simplified the Maykut and Untersteiner model and has obtained similar results with much less physical and numerical complexity.

Bryan et al. (1975) and Manabe et al. (1979) are examples of models that include crude parametrizations of ice motion and deformation. Hibler's (1979) model includes a more realistic treatment of the ice dynamics and its relation to thickness variations, but as for the Maykut and Untersteiner (1971) one, the numerical complexity of this model prevents its coupling with climate models.

Parkinson and Washington (1979) attempted to put an equivalent emphasis on thermodynamics and on dynamics, in a model that is compatible with climatic experiments. For the thermodynamics, they use a Semtner (1976)-

type of formulation, and the dynamics is based on the following five stresses : wind and water stresses, Coriolis force, internal ice resistance and the stress due to the tilt of the sea surface. A relatively good representation of the advance and retreat of the ice edge is obtained with this model both for the Arctic and the Antarctic, despite the lack of regard for conservation of momentum, as noted by Hibler (1980).

4. CATEGORIES OF SEA ICE EXPERIMENTS

The climate models used up to now for simulations involving sea ice can be classified into three broad categories, following Parkinson and Herman (1980) :

- the non-interactive sea ice models where sea ice just responds to the atmospheric forcing, without feeding it back. This allows sensitivity studies to be made, but a shortcoming is that the forcing may be inconsistent with the actual state of the ice.

- the non-interactive atmospheric models, where sea ice is prescribed as a lower boundary condition and remains independent of the model-generated fields. The lack of feedbacks between ice and atmosphere prevents the use without caution of the results of experiments with these models to study the effect of sea ice on the climate.

- the fully coupled sea ice-atmosphere models, which allow the distribution of ice to evolve as part of the solution to the complete system of equations governing atmospheric and sea ice processes. Coupled sea ice-ocean and sea-ocean-atmosphere models fall in this category also.

The preliminary results presented below were obtained with a non-interactive thermodynamic sea ice model forced with fields generated by a non interactive atmospheric model. These experiments are considered as a preparation for the development of a fully coupled sea ice-atmosphere model.

5. PRELIMINARY RESULTS

The one-dimensional thermodynamical sea ice model has been developed following Semtner (1976) assumption that for such a vertical model, thermodynamics plays the major role : no dynamical effects are taken into account, and the seasonal cycle of sea ice thickness is obtained as a result of bottom accretion and surface and bottom meltin due to local energy imbalances (cfr van Ypersele and Berger, 1982 for more details).

This model need as an input the following energy fluxes and date : the incident solar flux, the downward infrared atmospheric radiation and the latent and sensible heat fluxes between ice and atmosphere. All these fluxes are taken from the EERM GCM seasonal cycle experiment (cfr the paper by Roger on this GCM in this volume). The model needs also a value for the oceanic heat flux, taken here as a constant in the Arctic (= 2 W m^{-2}), the snowfall, which is also derived from the GCM output, and the annual cycle of snow albedo, taken from Maykut and Untersteiner (1971).

For each GCM grid point with latitudes between 65° and 90° N and for which the GCM prescribes the presence of sea ice, an integration of the model is made over a few years (we have only one year of GCM output, but we use the same yearly date again and again). The following figure shows the isolines of sea ice thickness for the months of January, April, July and October.

A preliminary conclusion that can be drawn from this experiment is that the simulated sea ice pattern and evolution are very similar to the ones that are assumed by the GCM : no more than four grid points where the GCM assumes the presence of ice are, according to this experiment, ice free at one season or another.

As it was mentioned in section 4, these results have to be examined with much caution, since there was no coupling between the GCM forcing and the actual state of the ice. Further work is definitely needed in that respect, and this experiment represented only a preparation for a future coupling.

ACKNOWLEDGEMENTS

This work would not have been possible without the very kind collaboration of Jean-François Royer (EERM, Météorologie Nationale, Toulouse), nor without the support of the Commission of the European Communities (contract n° CL-026-B(D)).

REFERENCES

Bryan, K., Manabe, S., Pacanowski, R.C. (1975), A global ocean-atmosphere climate model. Part II. The oceanic circulation. Journal of Physical Oceanography, vol. 5, pp. 30-46.

Budyko, M.I. (1974), Climate and Life. International Geophysics Series, vol. 18. Academic Press, New York, 508 pp.

Hibler III, W.D. (1979), A dynamic thermodynamic sea ice model. Journal of Physical Oceanography, vol. 9, n° 7, pp. 815-846.

Hibler III, W.D. (1980), Sea ice growth, drift, and decay, in "Dynamics of snow and ice masses", ed. by S.C. Colbeck, Academic Press, New York, pp. 141-209.

Manabe, S., Bryan, K., Spelman, M.J. (1979), A global ocean-atmosphere climate model with seasonal variation for future studies of climate sensitivity. Dyn. Atmos. Oceans, vol. 3, pp. 393-426.

Maykut, G.A. and Untersteiner, N. (1971), Some results from a time-dependent thermodynamic model of sea ice. Journal of Geophysical Research, vol. 76, n° 6, pp. 1550-1575.

Parkinson, C.L. and Herman, G.F. (1980), Sea ice simulations based on fields generated by the GLAS GCM. Monthly Weather Review, vol. 108, n° 12, pp. 2080-2091.

Parkinson, C.L. and Washington, W.M. (1979), A large scale numerical model of sea ice. Journal of Geophysical Research, vol. 84, Cl, pp. 311-337.

Semtner, A.J. Jr. (1976), A model for the thermodynamic growth of sea ice in numerical investigations of climate. Journal of Physical Oceanography, vol. 6, n° 3, pp. 379-389.

Untersteiner, N., (1975), Sea ice and ice sheets and their role in climatic variations, in "The physical basis of climate and climate modelling", GARP Publications series n° 16, W.M.O., Geneva, pp. 206-224.

van Ypersele, J.P. and Berger A. (1982), A coupled one-dimensional atmosphere - ocean -sea-ice model for climate studies, in "Extended abstracts and texts of presentations at the second meeting of the Contact Group "Climate Models", Brussels, 11th May 1982", Publication of the Commission of the European Communities n° XXII/CLI/3/82, pp. 35-39.

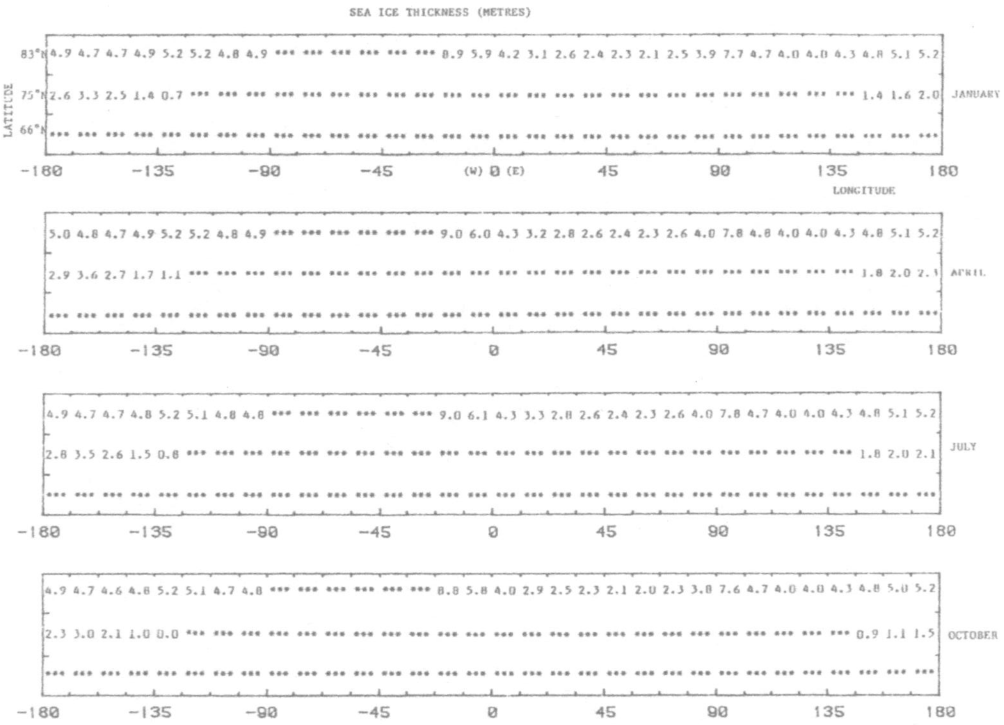

SEA ICE THICKNESS (METRES)

Figure 1 : Values of the sea ice thickness (m) in the northern polar latitudes, as simulated by a thermodynamical sea ice model forced with EBRM GCM energy fluxes for the months of January, April, July and October (see text for details). The asterisks represent points which, according to the boundary conditions of the GCM, are on continents or are not permanently ice-covered. There is a discrepancy between these results and the boundary conditions for one point only (and not four, as stated erroneously in the text).

LIST OF PARTICIPANTS

TITLE AND NAME   INSTITUTION AND ADDRESS

Prof. A.L. BERGER   Université Catholique de Louvain
                          Institut d'Astronomie et Géophysique
                          Chemin du Cyclotron, 2
                          1348 - LOUVAIN-LA-NEUVE
                          Belgium

Mr. V. CANTU'   Servizio Meteorologico dell'Aeronautica
                          militare
                          Aeroporto di VIGNA DI VALLE
                          00062 - BRACCIANO
                          Italy

Prof. W. DANSGAARD   Geofysik Isotoplaboratorium
                          Kobenhavns Universitet
                          Haraldsgade 6
                          2200 KOBENHAVN N
                          Denmark

Dr. J.C. DUPLESSY   Centre des Faibles Radioactivités
                          CNRS
                          91190 GIF-SUR-YVETTE
                          France

Dr. A. FERHI   Université Paris VI
                          Dépt. de Géographie Physique
                          4, Place Jussieu
                          75230 PARIS CEDEX 05
                          France

Prof. Dr. H. FLOHN   Meteorologisches Institut
                          der Universität Bonn
                          Auf dem Hügel 20
                          5300 BONN 1
                          F.R. Germany

Mr. Ph. GASPAR   Université Catholique de Louvain
                          Institut d'Astronomie et Géophysique
                          Chemin du Cyclotron, 2
                          1348 - LOUVAIN-LA-NEUVE
                          Belgium

Mr. H. GEURTS   KNMI
                          Wilhelminalaan 10
                          DE BILT
                          The Netherlands

Prof. K. HASSELMANN   Max-Planck-Institut für Meteorologie
                          Bundesstrasse 55
                          2000 HAMBURG 13
                          F.R. Germany

Mrs. Evang. M. HATZIOTIS   Service Archéologique de Grèce
Hippokratous 152
ATHENES (705)
Greece

Dr. G.P. HEKSTRA   Ministerie van Volkshuisvesting.
Ruimtelijke Ordening en Milieubeheer
Postbus 439
2260 AK LEIDSCHENDAM
The Netherlands

Mr. T.S. HILLS   Meteorological Office
London Road
BRACKNELL
Berkshire RG12 2SZ
United Kingdom

Mr. E. KALLEN   University of Utrecht
Princetonplein 5
UTRECHT
The Netherlands

Mr. H.J. KRIJNEN   KNMI
P.O. Box 201
3730 AE DE BILT
The Netherlands

Mrs. K. LAVAL   Laboratoire de Météorologie Dynamique
ENS
24 rue Lhomond
PARIS
France

Dr. J.F.B. MITCHELL   Meteorological Office
London Road
BRACKNELL
Berkshire RG12 2SZ
United Kingdom

Dr. N. MURDOCH   University of Exeter
Department of Mathematics
North Park Road
EXETER EX4 4QE
United Kingdom

Dr. W.D. NESTEROFF   Université Pierre et Marie Curie
Dépt. de Géologie Dynamique
4, Place Jussieu
75230 PARIS CEDEX 05
France

Prof. G. NICOLIS   Université Libre de Bruxelles
Faculté des Sciences
c.p. 220
BRUXELLES
Belgium

Prof. J. OERLEMANS   University of Utrecht
                           Princetonplein 5
                           UTRECHT
                           The Netherlands

Prof. H. OESCHGER   Physics Institute
                           University of Bern
                           Sidlerstrasse 5
                           3012 BERN
                           Switzerland

Prof. R. PAEPE   Vrije Universiteit Brussel
                           Kwartairgeologie
                           Pleinlaan 2
                           1050 -BRUSSEL
                           Belgium

Prof. A. PONS   Faculté des Sciences St. Jérôme
                           Lab. de Botanique historique et
                           Palynologie
                           U.D.E.S.A.M.
                           Rue H. Poincaré
                           13397 MARSEILLE CEDEX 4
                           France

Dr. C. PUJOL   Lab. Géologie et Océanographie
                           Université Bordeaux I
                           33400 TALENCE
                           France

Prof. P. ROGNON   Université Pierre et Marie Curie
                           Dépt. de Géographie Physique
                           4, Place Jussieu
                           75230 PARIS CEDEX 05
                           France

Dr. J.F. ROYER   C.N.R.M.
                           Avenue Eisenhower
                           31057 TOULOUSE CEDEX
                           France

Dr. G.H. SCHLESER   KFA
                           Institut für Chemie
                           Postfacht 1913
                           5170 JULICH 1
                           F.R. Germany

Prof. G. SERET   UCL
                           Unité de Géographie Physique et
                           de Géologie du Quaternaire
                           Bâtiment Mercator
                           Place L. Pasteur 3
                           1348 - LOUVAIN-LA-NEUVE
                           Belgium

Dr. N.J. SHACKLETON   Godwin Laboratory
                           University of Cambridge
                           Free School Lane
                           CAMBRIDGE CB2 3RS
                           United Kingdom

Dr. R. SNEYERS   Institut Royal Météorologique
                                de Belgique
                                Avenue Circulaire, 3
                                1180 - BRUXELLES
                                Belgium

Prof. J. THIEDE   Geol.-Pal. Institute
                                University of Kiel
                                Olshausenstrasse 40/60
                                2300 KIEL
                                F.R. Germany

Dr. J. THOREZ   Laboratoire des Argiles
                                Université de Liège
                                B6 - Sart Tilman
                                4000 - LIEGE
                                Belgium

Dr. A.F. VAN ENGELEN   KNMI
                                Wilhelminalaan 10
                                DE BILT
                                The Netherlands

Mrs. E. VAN OVERLOOP   Vrije Universiteit Brussel
                                Kwartairgeologie
                                Pleinlaan, 2
                                1050 - BRUSSEL
                                Belgium

Mr. J.-P. VAN YPERSELE   Institut d'Astronomie et de Géophysique
                                UCL
                                Chemin du Cyclotron, 2
                                1348 - LOUNVAIN-LA-NEUVE
                                Belgium

Prof. W. VAN ZEIST   Biologisch-Archaeologisch Instituut
                                Poststraat 6
                                9712 ER GRONINGEN
                                The Netherlands

Dr. W.A. WATTS   Trinity College
                                DUBLIN 2
                                Ireland

Commission

Dr. Ph. BOURDEAU
Dr. R. FANTECHI
Dr. A. GHAZI   DG XII/G
Dr. H. OTT
Miss S. LEROY
Mrs. M. VAN DER WERF

Appendix

WORKSHOP COMMITTEE

Programme Committee : A. BERGER
                      J. DUPLESSY
                      H. FLOHN
                      A. GHAZI
                      A. GILCHRIST

Rapporteurs :

Session A   :   W. DANSGAARD
                  R. PAEPE
                  W. WATTS

Session B   :   J. DUPLESSY
                  N. SCHACKLETON

Session C   :   J. MITCHELL
                  J. THIEDE

Apendix to the session reports :

                  J. ROYER